Android 6 Essentials

Design, build, and create your own applications using the full range of features available in Android 6

Yossi Elkrief

BIRMINGHAM - MUMBAI

Android 6 Essentials

First published: November 2015

Production reference: 1251115

Published by Packt Publishing Ltd.
Livery Place
35 Livery Street
Birmingham B3 2PB, UK.

ISBN 978-1-78588-441-2

www.packtpub.com

Credits

Author
Yossi Elkrief

Reviewer
Pavel Pavlasek

Commissioning Editor
Edward Gordon

Acquisition Editor
Reshma Raman

Content Development Editor
Riddhi Tuljapurkar

Technical Editor
Gaurav Suri

Copy Editor
Stuti Srivastava

Project Coordinator
Sanchita Mandal

Proofreader
Safis Editing

Indexer
Hemangini Bari

Graphics
Kirk D'Penha

Disha Haria

Abhinash Sahu

Production Coordinator
Aparna Bhagat

Cover Work
Aparna Bhagat

About the Author

Yossi Elkrief is an Android enthusiast with over 7 years of experience in the Android platform and is currently working as an Android architect and group leader at Tikal Knowledge.

Among his previous experiences, the noteworthy ones include ooVoo, Fiverr, Mobli, and Glide, reaching out to over 135 million users worldwide.

Yossi is a mentor at Google Launchpad, a lecturer on IoT and mobile development, and co-tech lead on the Madgera accelerator. He cofounded the GDG Be'er Sheva group and co-leads the group today, holding technology events for the technology community in Israel. He has a spouse and a daughter, and he lives in Be'er Sheva, Israel.

His main interests are Liverpool Football Club and his Android mini collectibles collection, with over 120 different pieces. You can find him on LinkedIn at `https://il.linkedin.com/in/yossielkrief`, on GitHub at `MaTriXy`, and on Google+ at Yossi.Elkrief.

Acknowledgments

First, I want to thank my family for their patience, love, and endurance with me taking yet another challenge, reducing the amount of time I can spend with them.

To Irit, my wife, for the constant love and support and remembering to feed me when I couldn't remember to feed myself.

To Mia, my daughter I love you.

To my mother and father, whom I love and cherish.

To my friends, for being there with their coding armor on and for giving me the chance to shine.

To my mentor, friend, and family, Israel Mali; thank you for guiding me for 15 years, helping me carve my career path. May you rest in peace.

I want to thank Packt Publishing for this opportunity and for publishing my first book. A special thanks to Riddhi Tuljapurkar for all the help and guidance.

Last but not least, I challenge you to try out the Chubby Bunny challenge at `http://icebreakerideas.com/chubby-bunny-challenge/`.

About the Reviewer

Pavel Pavlasek has been an Android developer for over 5 years, apart from being a long-time Java developer. He develops web information systems and Android applications, and he is passionate about new technologies.

I would like to thank to my family for their support and inspiration—my wife, Daniela, and my children, Michaela, Jakub, and Juraj. I would also like to thank my colleagues and other Android developers I've met.

www.PacktPub.com

Support files, eBooks, discount offers, and more

For support files and downloads related to your book, please visit www.PacktPub.com.

Did you know that Packt offers eBook versions of every book published, with PDF and ePub files available? You can upgrade to the eBook version at www.PacktPub.com and as a print book customer, you are entitled to a discount on the eBook copy. Get in touch with us at service@packtpub.com for more details.

At www.PacktPub.com, you can also read a collection of free technical articles, sign up for a range of free newsletters and receive exclusive discounts and offers on Packt books and eBooks.

https://www2.packtpub.com/books/subscription/packtlib

Do you need instant solutions to your IT questions? PacktLib is Packt's online digital book library. Here, you can search, access, and read Packt's entire library of books.

Why subscribe?

- Fully searchable across every book published by Packt
- Copy and paste, print, and bookmark content
- On demand and accessible via a web browser

Free access for Packt account holders

If you have an account with Packt at www.PacktPub.com, you can use this to access PacktLib today and view 9 entirely free books. Simply use your login credentials for immediate access.

Table of Contents

Preface   **v**

Chapter 1: Android Marshmallow Permissions   **1**

An overview of Android permissions   **1**
Permissions   2
Permission group definitions   3
Permissions that imply feature requirements   4
Viewing the permissions for each app   5
Understanding Android Marshmallow permissions   **9**
An overview   9
Permission groups   10
Runtime permissions   10
Taking coding permissions into account   **11**
Testing permissions   11
Coding for runtime permissions   12
Best practices and usage notes   13
Minimalism is a great option   14
Asking for too many permissions at once   14
Honesty can be a great policy   14
Need support handling runtime permissions?   **15**
Some permissions are normal and safer to use   17
Summary   **19**

Chapter 2: App Links   **21**

The Android Intent system   **21**
Creating a website association   23
Why this file?   23
Triggering app link verification   23
App link settings and management   24

Testing app links   **25**
  Checking manifest and listing domains   25
  The Digital Asset Links API   25
  Testing our intent   25
  Checking policies using adb   26
Summary   **28**

Chapter 3: Apps' Auto Backup   **29**
  An overview   **30**
  Data backup configuration   **30**
    Including or excluding data   31
    The backup configuration syntax   32
    Opting out from app data backup   33
  Backup configuration testing   **33**
    Setting backup logs   33
    Testing the backup phase   33
    Testing the restore phase   34
    Troubleshooting   34
  Important bytes   **35**
    What to exclude from the backup   36
    BackupAgent and backup events   36
  Summary   **37**

Chapter 4: Changes Unfold   **39**
  Power-saving modes   **40**
    The Doze mode   40
      What happens to apps when a device is dozing?   41
      Testing apps with Doze mode   41
    The App Standby mode   43
      What happens to apps when in the App Standby mode?   44
      Testing apps with the App Standby mode   44
      Excluded apps and settings   45
      Tips   48
  Removable storage adoption   **49**
  Apache HTTP client removal   **50**
  Notifications   **51**
  Text selection   **51**
    Support library notice   52
  Android Keystore changes   **52**
  Wi-Fi and networking changes   **52**
  Runtime   **53**
  Hardware identifier   **53**
  APK validation   **54**

USB connection	**54**
Direct Share	**54**
What if we have nothing to share?	57
Direct Share best practices	57
Voice interactions	**58**
The Assist API	**58**
Bluetooth API Changes	**59**
Bluetooth stylus support	59
Improved Bluetooth low energy scanning	60
Summary	**60**
Chapter 5: Audio, Video, and Camera Features	**61**
Audio features	**61**
Support for the MIDI protocol	61
MidiManager	62
Digital audio capture and playback	63
Audio and input devices	63
Information on audio devices	63
Changes in AudioManager	63
Video features	**64**
android.media.MediaSync	64
MediaCodecInfo.CodecCapabilities.getMaxSupportedInstances	64
Why do we need to know this?	64
MediaPlayer.setPlaybackParams	65
Camera features	**65**
The flashlight API	65
The reprocessing API	66
android.media.ImageWriter	66
android.media.ImageReader	66
Changes in the camera service	66
Summary	**67**
Chapter 6: Android for Work	**69**
Behavioral changes	**70**
The work contacts' display option	70
Wi-Fi configuration options	70
The Wi-Fi configuration lock	70
Work Policy Controller addition	71
DevicePolicyManager changes	71
Single-use device improvements	**72**
Silently installing/uninstalling apps	**73**
Improved certificate access	**73**
Automatic system updates	**73**

Third-party certificate installation   **74**
Data usage statistics   **74**
Managing runtime permissions   **74**
VPN access and display   **75**
Work profile status   **75**
Summary   **75**
Chapter 7: Chrome Custom Tabs   **77**
What is a Chrome custom tab?   **77**
  What is WebView?   77
  Customization options   78
When to use Chrome custom tabs   **78**
The implementation guide   **80**
  Can we use Chrome custom tabs?   80
  Custom UI and tab interaction   80
    The custom action button   83
    Configuring a custom menu   83
    Configuring custom enter and exit animations   84
    Chrome warm-up   84
    Connecting to the Chrome service   85
    Warming up the browser process   86
    Creating a new tab session   86
    Setting the prefetching URL   86
    Custom tabs connection callback   87
  Summary   **88**
Chapter 8: Authentication   **89**
The Fingerprint authentication API   **89**
  How do we use fingerprint authentication?   90
  Setting up for testing   91
Credentials' Grace Period   **93**
Cleartext network traffic   **94**
  So, what do we do with the cleartext network traffic flag? 94
Summary   **95**
Index   **97**

Preface

Android 6 will primarily focus on improving the overall user experience, and it will bring in a few features, such as a redesigned permission model in which applications are no longer automatically granted all of their specified permissions at the time of installation, the Doze power scheme for extended battery life when a device is not manipulated by the user, and native support for fingerprint recognition.

If you're already an Android developer, you're only a few steps away from being able to use your existing development experience to reach your users wherever or whenever they want or need your app.

As a professional Android developer, you have to create production-ready apps for your users. This book will give you what it takes to ship polished apps as part of a development team at a company, an independent app developer, or just as a programmer using Android development best practices.

By the end of the book, you'll be able to identify critical areas for improvement in an app and implement the necessary changes and refinements to ensure it meets Android's Core App Guidelines prior to shipping.

What this book covers

Chapter 1, *Android Marshmallow Permissions*, discusses how the Android permission system and model are vast and have made a few changes that can help app developers and applications gain more traction, installations, and give users the ability to decide when your applications will be able to use each permission-dependent feature. Keep in mind, though, that this is just a starting point and Android Marshmallow still needs to gain market share and get adopted by OEMs, enabling users with the freedom of choice. You as an app developer must prepare in advance and make sure your application development is forward-facing, allowing new users to enjoy the latest updates as soon as possible while maintaining a high level of performance for your applications.

Chapter 2, *App Links*, talks about how app linking has become powerful in Android Marshmallow. This allows you, the app developers, help the system better decide how to act. Handling web URLs will give you wider exposure, a bigger funnel into your apps, and better experience, which you can provide to your users (sums up to better ratings and more downloads and vice versa).

App linking is simple to implement, easy to understand, and is a must-have feature in the mobile/web world today. While app linking enables better action handling for users using your applications, users can have multiple devices, expecting the same behavior on each device, and would be more engaged if their data and action handling is all around.

Chapter 3, *Apps' Auto Backup*, informs you that Android Marshmallow brings with it a great backup feature for apps, reducing friction for users migrating to new devices.

In a world full of such diverse apps, maximizing the benefits from automatic backups leads to excellent user experience. The goal of this feature is to unload the burden and shorten the time required to set up a new device with the user's favorite apps. Allowing the users to enter your app with just a password prompt, if required, after a new installation can be a great experience. Try it yourself!

Chapter 4, *Changes Unfold*, goes over a few of the changes in Android Marshmallow. All of these changes are important to follow and will help you in your app development cycles. A few more changes are discussed in future chapters with a more detailed approach.

Chapter 5, *Audio, Video, and Camera Features*, covers quite a few changes and additions to Android APIs. Android Marshmallow is more about helping us, the developers, achieve better media support and showcase our ideas when using audio, video, or camera APIs.

Chapter 6, *Android for Work*, covers how Android Marshmallow has brought in quite a few changes to the world of Android for Work. As developers, we need to always maintain a viable connection with the needs of an organization. Making sure that we go over and understand the Android for Work world with the changes in Marshmallow helps us build and target enterprise workflows with the added benefit of a simpler API.

Chapter 7, *Chrome Custom Tabs*, talks about the newly added feature, Chrome custom tabs, that allows us developers to embed web content into our application, modify the UI, and adjust it to our app's theme and colors and the look and feel. This helps us keep the user in our application and still provide a nice UI and overall feel.

Chapter 8, Authentication, discuss how Android Marshmallow gives us a new API to authenticate users with the fingerprint API. We can use the sensor and authenticate the user even within our application and save it for later usage if we want to save the need of user login using the Credentials grace period abilities Android Marshmallow has introduced. We also covered a way to make our application more secure using HTTPS only. The StrictMode policy, enforced with the help of the usesCleartextTraffic flag, allows us to make sure that all the nodes we connect to the outer world are examined to check if there's a need for a secure connection or not.

What you need for this book

For this book, you will require previous knowledge of the Android platform, APIs, and the application development process. You will also need to set up your work environment to have at least the following:

- Android Studio, which can be downloaded from `https://developer.android.com/sdk/index.html`

- The latest Android SDK tools and platforms. Make sure that you upgrade to the latest versions and add the Android 6.0 (Marshmallow) platform if it's missing

- An Android device is helpful, but you may use an emulator if you prefer, or you may use the great solution of Genymotion as an emulator, at `https://www.genymotion.com/`

Who this book is for

This book is for Android developers who are looking to move their applications into the next Android version with ease. In the chapters of this book, the author has referred to Android 6 as Android Marshmallow. You should have a good understanding of Java and previous Android APIs, and you should be able to write applications with APIs prior to Marshmallow.

Conventions

In this book, you will find a number of text styles that distinguish between different kinds of information. Here are some examples of these styles and an explanation of their meaning.

Code words in text, database table names, folder names, filenames, file extensions, pathnames, dummy URLs, user input, and Twitter handles are shown as follows: "The `setTorchMode()` method has been added to control the flash torch mode."

A block of code is set as follows:

```xml
<?xml version="1.0" encoding="utf-8"?>
<full-backup-content>
  <exclude domain="database" path="sensitive_database_name.db"/>
  <exclude domain="sharedpref"
    path="androidapp_shared_prefs_name"/>
  <exclude domain="file" path="some_file.file_Extension"/>
  <exclude domain="file" path="some_file.file_Extension"/>
</full-backup-content>
```

Any command-line input or output is written as follows:

```
$ adb shell sm set-force-adoptable true
```

New terms and **important words** are shown in bold. Words that you see on the screen, for example, in menus or dialog boxes, appear in the text like this: "When heading to **Settings** | **More** | **VPN**, you can now view the VPN apps."

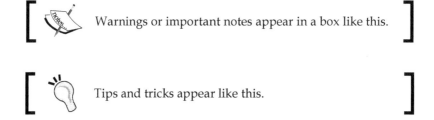

Warnings or important notes appear in a box like this.

Tips and tricks appear like this.

Reader feedback

Feedback from our readers is always welcome. Let us know what you think about this book—what you liked or disliked. Reader feedback is important for us as it helps us develop titles that you will really get the most out of.

To send us general feedback, simply e-mail feedback@packtpub.com, and mention the book's title in the subject of your message.

If there is a topic that you have expertise in and you are interested in either writing or contributing to a book, see our author guide at www.packtpub.com/authors.

Customer support

Now that you are the proud owner of a Packt book, we have a number of things to help you to get the most from your purchase.

Downloading the example code

You can download the example code files from your account at `http://www.packtpub.com` for all the Packt Publishing books you have purchased. If you purchased this book elsewhere, you can visit `http://www.packtpub.com/support` and register to have the files e-mailed directly to you.

Downloading the color images of this book

We also provide you with a PDF file that has color images of the screenshots/diagrams used in this book. The color images will help you better understand the changes in the output. You can download this file from `https://www.packtpub.com/sites/default/files/downloads/4412OS_ColoredImages.pdf`.

Errata

Although we have taken every care to ensure the accuracy of our content, mistakes do happen. If you find a mistake in one of our books—maybe a mistake in the text or the code—we would be grateful if you could report this to us. By doing so, you can save other readers from frustration and help us improve subsequent versions of this book. If you find any errata, please report them by visiting `http://www.packtpub.com/submit-errata`, selecting your book, clicking on the **Errata Submission Form** link, and entering the details of your errata. Once your errata are verified, your submission will be accepted and the errata will be uploaded to our website or added to any list of existing errata under the Errata section of that title.

To view the previously submitted errata, go to `https://www.packtpub.com/books/content/support` and enter the name of the book in the search field. The required information will appear under the **Errata** section.

Piracy

Piracy of copyrighted material on the Internet is an ongoing problem across all media. At Packt, we take the protection of our copyright and licenses very seriously. If you come across any illegal copies of our works in any form on the Internet, please provide us with the location address or website name immediately so that we can pursue a remedy.

Please contact us at `copyright@packtpub.com` with a link to the suspected pirated material.

We appreciate your help in protecting our authors and our ability to bring you valuable content.

Questions

If you have a problem with any aspect of this book, you can contact us at
questions@packtpub.com, and we will do our best to address the problem.

1

Android Marshmallow Permissions

Android permissions have been there for as long as we can remember — since Android 1.0, to be exact. Through the years and with the evolvement of platforms, the Android permissions model has been modified by adding new permissions and trying to allow more granular control over the part of the device hardware/data the application has.

In this chapter, we will review a bit of the Android permissions model that was prior to **Android Marshmallow**, and we'll focus on the changes it brings to the table. We will also explain the changes that you as a developer must do in order to handle all the other changes and make sure your applications work as intended on Android Marshmallow.

In this chapter, we will cover the following:

- An overview of Android permissions
- Understanding Android Marshmallow permissions
- Handling code permissions with best practices

An overview of Android permissions

In Android, each application runs with distinct system IDs known as **Linux user ID** and **Group ID**. The system parts are also separated into distinct IDs, forming isolated zones for applications — from each other and from the system. As part of this isolated life cycle scheme, accessing services or other applications' data requires that you declare this desire in advance by requesting a permission.

This is done by adding the `uses-permission` element to your `AndroidManifest.xml` file. Your manifest may have zero or more `uses-permission` elements, and all of them must be the direct children of the root `<manifest>` element.

Trying to access data or features without proper permission would give out a security exception (using a `SecurityException` class), informing us about the missing permission in most cases.

The `sendBroadcast(Intent)` method is exceptional as it checks permissions after the method call has returned, so we will not receive an exception if there are permission failures. A **permission failure** should be printed to the system log. Note that in Android versions prior to Marshmallow, missing permissions were due to missing declarations in the manifest. Hence, it is important that you keep permissions in mind when you come up with the feature list for your app.

Permissions

When using Android platform as an app, you have restrictions preventing access to some hardware, system APIs, private user data, and application data.

Permission is needed in order to allow access to a specific API, data, or hardware; it was asked upon the installation of your app up until Android Marshmallow. Most permissions are used to restrict access. When a permission is granted, you then have access to that specific restricted area. A feature can be protected by one permission at most.

The `uses-permission` element takes a name attribute, `android:name`, which is the name of the permission your application requires:

```
<uses-permission android:name="string"
  android:maxSdkVersion="integer" />
```

Did you know that the `android:maxSdkVersion` attribute, added in API level 19, is used to notify the version of the API from which this permission *should not* be granted? This is useful if a permission is no longer needed on higher versions of the API. For example, take a look at the following:

```
<uses-permission
  android:name="android.permission.READ_EXTERNAL_STORAGE"
  android:maxSdkVersion="18" />
```

In API 19, your app doesn't need to ask for this permission—it's granted to you.

Your application can also protect its own components, such as activities, services, broadcast receivers, and content providers with permissions.

It can employ any of the permissions defined by Android and declared by other applications, or it can define its own.

For more information on permissions, you can read `http://developer.android.com/reference/android/Manifest.permission.html`.

Permission group definitions

Permissions are divided into groups. According to Google, we can say that a **permission group** puts together related permissions in a single name/tag. You can group permissions together using the `permissionGroup` attribute inside the `<permission>` element.

Permissions grouped in the same permission group are shown as one group when approving permissions or when checking an app for its permissions.

The permission group is what you see when installing an application from the Play Store; for example, take a look at the following screenshot:

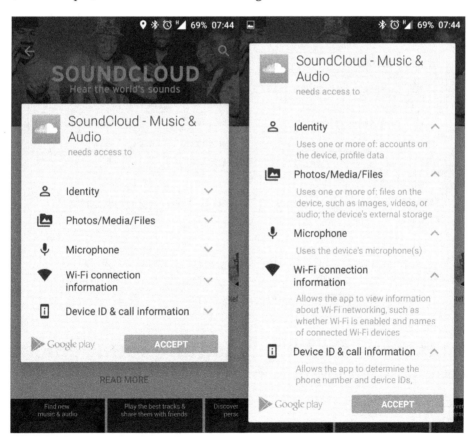

Let's take a look at the structure of the `permission-group` tag:

```
<permission-group android:description="string resource"
                  android:icon="drawable resource"
                  android:label="string resource"
                  android:name="string" />
```

The elements of the preceding structure can be explained as follows:

- `android:description`: This refers to simple text used to describe the group.
- `android:icon`: This refers to an icon from a drawable resource that represents the permission.
- `android:label`: This refers to a simple text name for the group.
- `android:name`: This is the name of the group. It is used to assign permissions to a specific group.

The following table shows you the various categories of permissions that are there in a permissions group:

Permissions group	
In-app purchases	Device and app history
Contacts	Calendar
Phone	Photos, media, and files
Wi-Fi connection information	Bluetooth connection information
Identity	Cellular data settings
SMS	Location
Microphone	Camera
Device ID and call information	Wearable sensors/activity data
Other	

Any permissions that are not part of a permissions group will be shown as **Other**. When an app is updated, there may be changes to the permissions group for that app.

Permissions that imply feature requirements

Some permissions are implied by feature requirements; we will cover this next.

When declaring a feature in the manifest, we must also request the permissions that we need.

Let's say, for example, that we want to have a feature that sets pictures for our contacts. If we want to take a picture via the `Camera` API, then we must request a `Camera` permission.

The `<users-feature>` tag makes sure we declare that we need devices that support the required feature for our application to work and use that feature. If this feature is not a required feature and our app can work without it but with fewer features, we can use `android:required="false"`, keeping it in mind that this feature is optional.

The `<uses-feature>` declarations always take precedence over features implied by permissions. The complete list of permission categories that imply feature requirements can be found at `http://developer.android.com/guide/topics/manifest/uses-feature-element.html#permissions`.

Viewing the permissions for each app

You can look at the permissions for each app using the settings app or the `adb` shell command.

To use the settings app, go to **Settings | Apps**. Pick an app and scroll down to see the permissions that the app uses You can see the Lollipop version in the following screenshot:

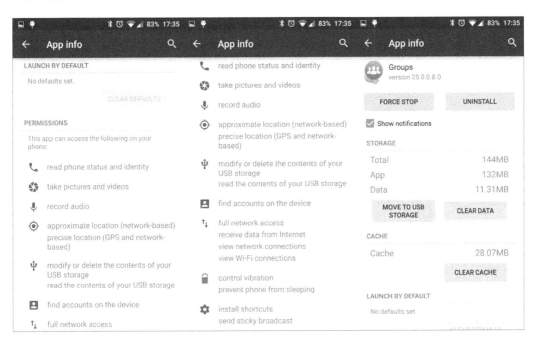

In Android Marshmallow, the UI is different.

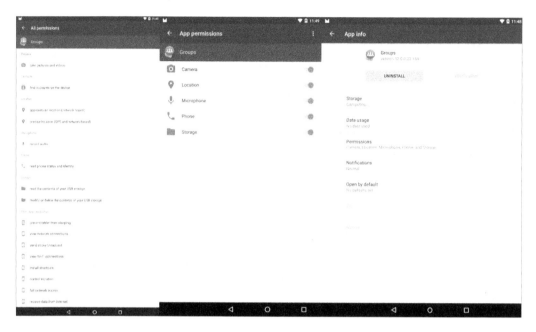

The second option is to use the `adb` shell commands with the `aapt` command:

1. List all the applications along with their installation paths. As an example, let's try to find out Facebook groups' app permissions using the following command:

   ```
   adb shell pm list packages -f
   ```

   We can use the `-3` flag to just show the third-party apps instead of the entire list.

2. Once we get the package location (`apk`), we need to pull it from the device via the `adb` pull:

   ```
   adb pull /data/app/com.facebook.groups-1/base.apk
   ```

3. Our final step to show permissions is to use `aapt` found in the `build-tools` folder of your specific build tools version:

```
aapt d permissions base.apk
```

This gives us the following screenshot as a result:

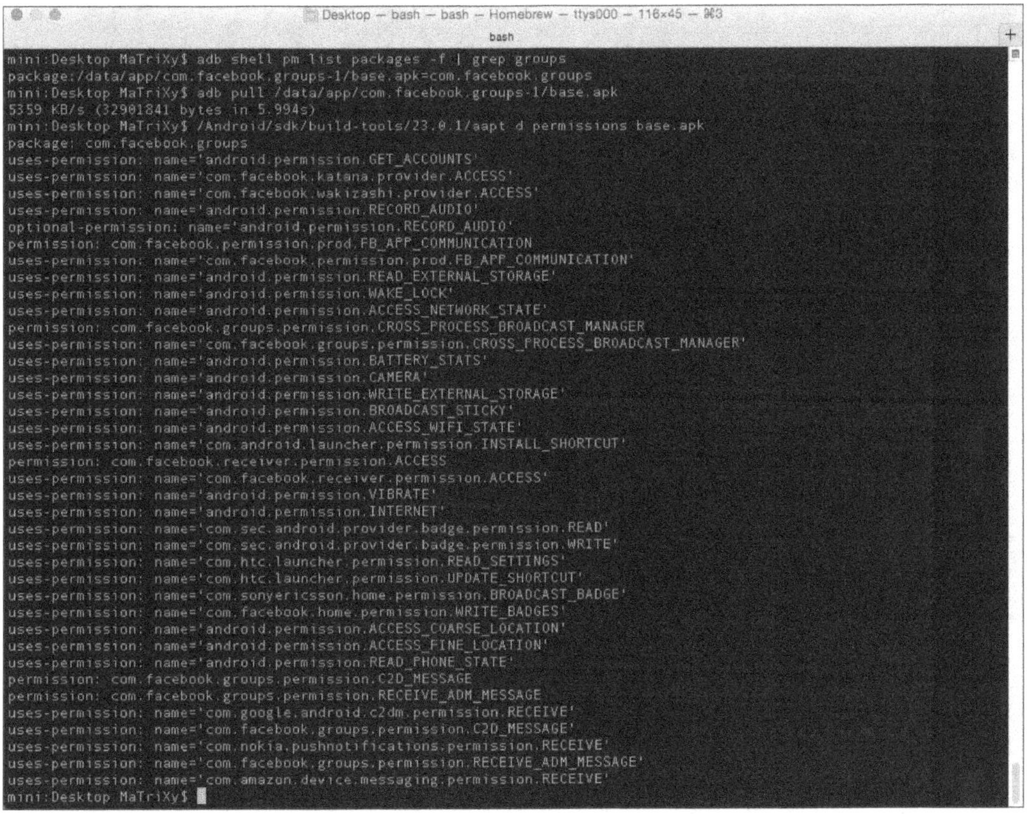

To view the permissions for the entire device, take a look at the following screenshot:

Using an `adb` command, you can print all known permissions on the device. The package manager (`pm`) command inside the `adb` command looks something like the following:

```
$ adb shell pm list permissions [options] <GROUP>
```

List permissions get the `[options]` and `<GROUP>` arguments (both optional).

Here, `options` can be as follows:

- `-g`: This refers to a list of permissions organized by a group
- `-f`: This prints all the information
- `-s`: This prints a short summary, and this is what the user sees on screen when checking permissions or approving them
- `-d`: This looks up and prints only permissions that are considered dangerous
- `-u`: This lists permissions visible to the user only

Understanding Android Marshmallow permissions

Android Marshmallow introduces a new application permissions model, allowing a simpler process for users when installing and/or upgrading applications. Applications running on Marshmallow should work according to a new permissions model, where the user can grant or revoke permissions after the installation—permissions are not given until there is user acceptance.

Supporting the new permissions model is backward-compatible, which means your apps can still be installed and run on devices running older versions of Android using the old permissions model on those devices.

An overview

With the Android Marshmallow version, a new application permissions model has been introduced.

Let's review it a bit more thoroughly:

- **Declaring permissions**: All permissions an app needs are declared in the manifest, which is done to preserve backward compatibility in a manner similar to earlier Android platform versions.

- **Permission groups**: As discussed previously, permissions are divided into permission groups based on their functionalities:

    - **PROTECTION_NORMAL permissions**: Some of the permissions are granted when users install the app. Upon installation, the system checks your app's manifest and automatically grants permissions that match the `PROTECTION_NORMAL` group.

    - **INTERNET permission**: One important permission is the `INTERNET` permission, which will be granted upon installation, and the user can't revoke it.

- **App signature permissions granted**: The user is not prompted to grant any permissions at the time of installation.

- **Permissions granted by users at runtime**: You as an app developer need to request a permission in your app; a system dialog is shown to the user, and the user response is passed back to your app, notifying whether the permission is granted.

- **Permissions can be revoked**: Users can revoke permissions that were granted previously. We must learn how to handle these cases, as we'll learn later on.

 If an app targets an Android Marshmallow version, it must use the new permissions model.

Permission groups

When working with permissions, we divide them into groups. This division is done for fast user interaction when reviewing and approving permissions. Granting is done only once per permission group. If you add a new permission or request a new permission from the same permission group and the user has already approved that group, the system will grant you the added permission without bothering the user about the approval.

For more information on this, visit `https://developer.android.com/reference/android/content/pm/PermissionInfo.html#constants`.

When the user installs an app, the app is granted only those permissions that are listed in the manifest that belongs to the `PROTECTION_NORMAL` group.

Requesting permissions from the `PROTECTION_SIGNATURE` group will be granted only if the application is signed with the same certificate as the app with the declared permission.

 Apps cannot request signature permissions at runtime.

System components automatically receive all the permissions listed in their manifests.

Runtime permissions

Android Marshmallow showcased a new permissions model where users were able to directly manage app permissions at application runtime. Google has altered the old permissions model, mostly to enable easier and frictionless installations and auto-updates for users as well as for app developers. This allows users to install the app without the need to preapprove each permission the application needs. The user can install the app without going through the phase of checking each permission and declining the installation due to a single permission.

Users can grant or revoke permissions for installed apps, leaving the tweaking and the freedom of choice in the users' hands.

Most of the applications will need to address these issues when updating the target API to 23.

Taking coding permissions into account

Well, after all the explanations, we've reached the coding part, and this is where we will get our coding hands dirty. The following are key methods used for handling permissions:

- `Context.checkSelfPermission()`: This checks whether your app has been granted a permission
- `Activity.requestPermission()`: This requests a permission at runtime

Even if your app is not yet targeting Android Marshmallow, you should test your app and prepare to support it.

Testing permissions

In the Android Marshmallow permissions model, your app must ask the user for individual permissions at runtime. There is limited compatibility support for legacy apps, and you should test your app and also test a version to make sure it's supported.

You can use the following test guide and conduct app testing with the new behavior:

- Map your app's permissions
- Test flows with permissions granted and revoked

The `adb` command shell can be quite helpful to check for permissions:

- Listing application permissions and status by group can be done using the following `adb` command:

  ```
  adb shell pm list permissions -g
  ```

- You can grant or revoke permissions using the following `adb` syntax:

  ```
  adb shell pm [grant|revoke] <permission.name>
  ```

- You can grant permissions and install `apk` using the following `adb` command:

  ```
  adb install -g <path_to_apk>
  ```

Coding for runtime permissions

When we want to adjust our application to the new model, we need to make sure that we organize our steps and leave no permission stranded:

- **Check what platform the app is running on**: When running a piece of code that is sensitive at the API level, we start by checking the version/API level that we are running on.

  By now, you should be familiar with `Build.VERSION.SDK_INT`.

- **Check whether the app has the required permission**: Here, we get ourselves a brand new API call:

  `Context.checkSelfPermission(String permission_name)`.

  With this, we silently check whether permissions are granted or not.

  This method returns immediately, so any permission-related controls/flows should be dealt with by checking this first.

- **Prompting for permissions**: We have a new API call, `Activity.requestPermissions (String[] permissions, int requestCode)`. This call triggers the system to show the dialog requesting a permission. This method functions asynchronously.

  You can request more than one permission at once. The second argument is a simple request code returned in the callback so that you can recognize the calls. This is just like how we've been dealing with `startActivityForResult()` and `onActivityResult()` for years.

  Another new API is `Activity.shouldShowRequestPermissionRationale( String permission)`.

  This method returns `true` when you have requested a permission and the user denied the request. It's considered a good practice after verifying that you explain to the user why you need that exact permission. The user can decide to turn down the permission request and select the *Don't ask again* option; then, this method will return `false`.

The following sample code checks whether the app has permission to read the user's contacts. It requests the permission if required, and the result callback returns to `onRequestPermissionsResult`:

```
if (checkSelfPermission(Manifest.permission.READ_CONTACTS) !=
  PackageManager.PERMISSION_GRANTED) {
  requestPermissions(new
  String[]{Manifest.permission.READ_CONTACTS},
  SAMPLE_MATRIXY_READ_CONTACTS);
}
```

```
//Now this is our callback
@Override
public void onRequestPermissionsResult(int requestCode, String
  permissions[], int[] grantResults) {
  switch (requestCode) {
  case SAMPLE_MATRIXY_READ_CONTACTS:
    if (grantResults[0] == PackageManager.PERMISSION_GRANTED) {
      // permission granted - we can continue the feature flow.
    } else {
      // permission denied! - we should disable the functionality
      that depends on this permission.
    }
  }
}
```

Just to make sure we all know the constants used, here's the explanation:

- `public static final int PERMISSION_DENIED=-1`:

  Since it's API level 1, permission has not been granted to the given package

- `public static final int PERMISSION_GRANTED=0`:

  Since it's API level 1, permission has been granted to the given package.

If the user denies your permission request, your app should take the appropriate action, such as notifying the user why this permission is required or explaining that the feature can't work without it.

Your app cannot assume user interaction has taken place because the user can choose to reject granting a permission along with the *do not show again* option; your permission request is automatically rejected and `onRequestPermissionsResult` gets the result back.

Best practices and usage notes

The new permissions model has brought to life a smoother experience for users and a bit more code-handling for developers. It makes it easier to install and update apps and feel comfortable with what the apps are doing.

Minimalism is a great option

Don't be a permission hog! In our application life cycle, we should try to minimize our permission requests. Asking for a lot of permissions and maintaining them can seem hazardous for some, and we should try and make the feature smooth and ask for the smallest number of permissions as far as possible in order to allow relaxed, undisturbed usage. Consider using intents whenever possible—rely on other applications doing some of the work for us (fewer permissions means less friction, turning a good app into a great one).

Asking for too many permissions at once

Users can get distracted by too many dialogs popping up, asking them for more and more permissions. Instead, you should ask for permissions as and when you need them.

However, we have some exceptions to every rule. Your app may require a few permissions to begin with, such as a camera application showing the camera permissions right at the beginning. However, setting the photo to your contact can be done and requested only when the user triggers that specific action. Try to map your flow and make it easier for users to understand what is going on. Users will understand that you've requested permissions for contacts if they have asked to set information to a contact via your app.

One more suggestion: apps with a tutorial can integrate the essential permissions' request in the tutorial, allowing the users to better understand the flow and why each permission is used.

Honesty can be a great policy

When asking for a permission, the system shows a dialog stating which permission your app wants, but it doesn't say why. Consider users who hate being left in the dark thinking why this permission is needed now or users who deny the permissions due to speculation. Things can be even worse: sometimes, a user's cursor may be 2 cm away from the 1-star rating or the uninstall button.

This is why it's a good idea to explain why your app wants the permissions before calling `requestPermissions()`.

Keep in mind that most developers will choose a tutorial but a lot of users may choose to skip tutorials whenever possible, so you must make sure that you can provide information about permissions, apart from the ones in the tutorial.

Need support handling runtime permissions?

Managing permissions is easier with the latest revision of the **v4** or **v13** support libraries (23, which is the same as the Android Marshmallow API version, so it's easy to remember)

The support libraries now provide several new methods to manage permissions and work properly on any device that can use these libraries. This, for instance, saves you the time required to check for a sufficient API level regardless of whether the device runs Android Marshmallow or not. If an app is installed on a device running Android Marshmallow, proper behavior is achieved — as if you're running the same framework calls. Even when running on lower versions, you get the expected behavior from the support library methods.

The v4 support library has the following methods:

- `ActivityCompat.checkSelfPermission (Context context, String permission)`:

  This checks whether your app has a permission. `PERMISSION_GRANTED` is returned if the app has the permission; otherwise, `PERMISSION_DENIED` is returned.

- `ActivityCompat.requestPermissions (Activity activity, String[] permissions, int requestCode`:

  This requests permissions, if required. If the device is not running Android 6.0, you will get a callback.

- `ActivityCompat.OnRequestPermissionsResultCallback(int requestCode, String[] permissions, int[] grantResults)`:

  This passes `PERMISSION_GRANTED` if the app already has the specified permission and `PERMISSION_DENIED` if it does not.

- `ActivityCompat.shouldShowRequestPermissionRationale (Activity activity, String permission)`:

  This returns `true` if the user has denied a permission request at least once and has not yet selected the *Don't ask again* option.

According to the design patterns, we should now give our users more information about the feature and why these permissions are so important to the app.

If the device is not running Android Marshmallow, shouldShowRequestPermissionRationale will always return false.

The PermissionChecker class is also included in v4.

This class provides several methods for apps that use IPC to check whether a particular package has a specified permission when IPC calls are made.

Android has a compatibility mode, allowing users to revoke access to permission-protected methods for legacy apps. When a user revokes access in the compatibility mode, the app's permissions remain the same but access to the APIs is restricted.

The PermissionChecker method verifies app permissions in normal as well as legacy modes.

 If your app acts as a middleman on behalf of other apps and needs to call platform methods that require runtime permissions, you should use the appropriate PermissionChecker method in order to ensure that the other app is allowed to perform the operation.

The v13 support library provides the following permission methods:

- FragmentCompat.requestPermissions():

  This requests permissions, if required. If the device is not running Android 6.0, you will get a callback.

- FragmentCompat.OnRequestPermissionsResultCallback:

  This passes PERMISSION_GRANTED if the app already has the specified permission and PERMISSION_DENIED if it does not.

- FragmentCompat.shouldShowRequestPermissionRationale():

  This returns true if the user has denied a permission request at least once and has not yet selected the *Don't ask again* option.

According to the design patterns, we should now give our users more information about the feature and why this permission is so important to the app.

 If the device is not running Android Marshmallow, it will always return false.

You can check out the sample project for the three ways to handle permissions:

`https://github.com/MaTriXy/PermissionMigrationGuide`

For more information on permission design patterns, read *Patterns – Permissions* by Google at `https://www.google.com/design/spec/patterns/permissions.html`.

Some permissions are normal and safer to use

The Android system flags permissions according to their protection levels. The levels are describes at `http://developer.android.com/reference/android/content/pm/PermissionInfo.html`.

The level that is relevant to our discussion is `PROTECTION_NORMAL`, in which permissions are considered to have little or no risk when applications have them.

Let's say you want to build a flashlight app; allowing your app to turn on the flash is not considered a huge risk to privacy or security, and this is why flashlight permission is flagged `PROTECTION_NORMAL`.

When you declare normal permissions in the manifest, the system grants these permissions automatically at the time of installation. There is no prompt to grant permissions for a normal permissions group, and these permissions can't be revoked by users.

This means that you can be sure that normal permissions are granted at the time of installation.

Currently, the permissions classified as `PROTECTION_NORMAL` are as follows:

- `android.permission.ACCESS_LOCATION_EXTRA_COMMANDS`
- `android.permission.ACCESS_NETWORK_STATE`
- `android.permission.ACCESS_WIFI_STATE`
- `android.permission.ACCESS_WIMAX_STATE`
- `android.permission.BLUETOOTH`
- `android.permission.BLUETOOTH_ADMIN`
- `android.permission.BROADCAST_STICKY`
- `android.permission.CHANGE_NETWORK_STATE`
- `android.permission.CHANGE_WIFI_MULTICAST_STATE`

- android.permission.CHANGE_WIFI_STATE
- android.permission.DISABLE_KEYGUARD
- android.permission.EXPAND_STATUS_BAR
- android.permission.FLASHLIGHT
- android.permission.GET_ACCOUNTS
- android.permission.GET_PACKAGE_SIZE
- android.permission.INTERNET
- android.permission.KILL_BACKGROUND_PROCESSES
- android.permission.MODIFY_AUDIO_SETTINGS
- android.permission.NFC
- android.permission.PERSISTENT_ACTIVITY
- android.permission.READ_SYNC_SETTINGS
- android.permission.READ_SYNC_STATS
- android.permission.READ_USER_DICTIONARY
- android.permission.RECEIVE_BOOT_COMPLETED
- android.permission.REORDER_TASKS
- android.permission.SET_TIME_ZONE
- android.permission.SET_WALLPAPER
- android.permission.SET_WALLPAPER_HINTS
- android.permission.SUBSCRIBED_FEEDS_READ
- android.permission.TRANSMIT_IR
- android.permission.VIBRATE
- android.permission.WAKE_LOCK
- android.permission.WRITE_SETTINGS
- android.permission.WRITE_SYNC_SETTINGS
- android.permission.WRITE_USER_DICTIONARY
- com.android.alarm.permission.SET_ALARM
- com.android.launcher.permission.INSTALL_SHORTCUT

Summary

As you saw, the Android permission system and model is vast and has introduced a few changes that can help app developers and applications gain more traction and installations and give the users the ability to decide when your applications will be able to use each permission-dependent feature. Keep in mind, though, that this is just a starting point and Android Marshmallow still needs to gain market share and get adopted by OEMs, enabling users with freedom of choice. You as an app developer must prepare in advance and make sure your application development is forward-facing, allowing new users to enjoy the latest updates as soon as possible while maintaining a high level of performance for your applications.

In the next chapter, we will go over a small yet important feature in the Android Marshmallow version: app linking.

2
App Links

One of the major improvements to the new Android Marshmallow version is powerful **app linking**. It allows the association of your app with your owned web domain. With this association, you as a developer allow the system to determine the default app that should handle a particular web link and skip prompting users to select an app. Saving clicks means less friction, which means that you reach the content faster; this leads to users and developers being happy. In this chapter, we will cover the following topics:

- The Android Intent system
- Creating a website association
- Triggering app link verification
- App link settings and management
- Testing app links

The Android Intent system

Almost every developer knows what an **Android Intent** system is, but we will explain it a bit and lay out the basic principles required to understand the app links feature. The Android Intent system can be found in the Android platform; this allows the passing of data in a small, simple package. Intent means that we want to perform an action. You may already know about the basic intents:

- `startActivity()`
- `startActivityForResult()`
- `startService()`
- `sendBroadcast()`

The following figure shows an Android Intent system for the `startActivity()` and `onCreate()` intents:

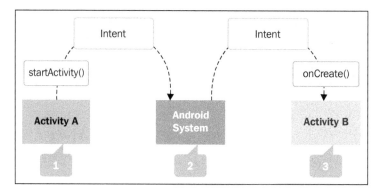

Source: http://developer.android.com/guide/components/intents-filters.html

The Android Intent system is a flexible mechanism that is used to enable apps to handle content and requests. Multiple apps may declare matching **URI** (short for **Uniform Resource Identifier**) patterns in their **intent filters**. When a user clicks on a web link that does not have a default **launch handler**, the platform may display a dialog for the user to select from a list of apps that have declared matching intent filters.

Intents are also used to trigger actions across the system, and some of these actions are system-defined, such as `ACTION_SEND` (referred to as the *share* action), where you as an app developer can share/send specific information to another application in order to complete an action required by a user.

Until Android Marshmallow, browsers handled each link clicked on the Web, and the system checked whether a **custom URI scheme** was available. Your application could handle specific custom actions via the custom URI scheme. This was tricky at times and didn't allow the handling of links under an entire web domain. Now, it's possible. Android Marshmallow's added support for app links allows you, as an app developer, to associate an app with a web domain. Automatically, this will allow you to determine the default app that will handle a particular web link instead of showing the selected application to handle the dialog.

 If you wish to read more about intents, you can go to the following link: http://developer.android.com/guide/components/intents-filters.html

Creating a website association

You as an app developer as well as a website owner need to declare a website association with an app. The declaration is done by hosting a JSON file, which is specifically named `assetlinks.json`. The file must reside in a specific location on the domain in question, such as:

```
https://<domain>:<optional port>/.well-known/assetlinks.json
```

 This file is accessed and verified over the HTTPS protocol and not HTTP.

Why this file?

The JSON file holds information about the Android application that will be the default handler for the URLs under this domain. In the JSON file, you must have the following structure:

```
[{
  "relation": ["delegate_permission/common.handle_all_urls"],
  "target": {
    "namespace": "android_app",
    "package_name": "com.yourapp.androidapp",
    "sha256_cert_fingerprints": [""]
  }
}]
```

The following are some elements of the preceding structure:

*   `package_name`: This is the package name from your app's **manifest**
*   `sha256_cert_fingerprints`: This is the SHA-256 fingerprint of your app

    Use the following command if you don't have this **SHA** (short for **Secure Hash Algorithm**):

    ```
    keytool -list -v -keystore app_release_signing.keystore
    ```

Triggering app link verification

You can request automatic verification for any app links declared in the `assetlinks.json` file. Requesting a verification is done by adding the `android:autoVerify` attribute to each intent filter in the manifest and setting it to `true`.

Let's say we own a WhatsApp application and domain. We want to autoverify an intent filter that has the `android.intent.action.VIEW` action.

The following is a sample activity from WhatsApp that handles app links and the autoverification attribute:

```
<activity android:name="com.whatsapp.XXX" …>
  <intent-filter android:autoVerify="true">
    <action android:name="android.intent.action.VIEW"/>
    <category android:name="android.intent.category.DEFAULT"/>
    <category android:name="android.intent.category.BROWSABLE"/>
    <data android:scheme="http" android:host="www.whatsapp.com"/>
    <data android:scheme="https" android:host="www.whatsapp.com"/>
  </intent-filter>
</activity>
```

The `android:autoVerify` attribute alerts the platform to verify app links when the app is installed. If the app link's verification fails, your app will not be set as the preferred app to handle these links. If there is no preferred app to handle these links whenever a user opens one of them, a dialog to choose an app is displayed.

If a user has used the system settings and set an app as the preferred app, then the link will go directly to the app but not because the verification was successful.

App link settings and management

For easy management, you can enter the system settings and tweak the URL handling by navigating to **Settings | Apps | App info | Open by default**.

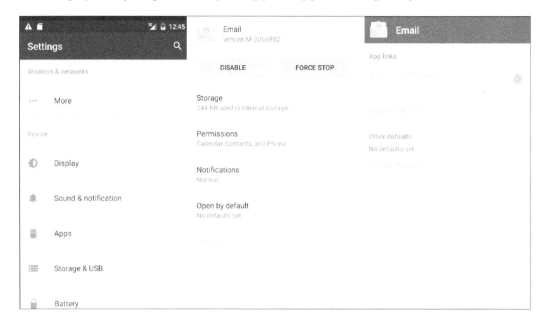

Testing app links

As with every new feature we add, we must test the app links feature that we will add to our application.

Checking manifest and listing domains

Our first step is to go over the manifest and make sure all the domains are registered correctly and all intent filters are well defined. Only the links/domains under all the criteria mentioned in the following bullets are the ones we need to test:

- The `android:scheme` attribute with a value of HTTP or HTTPS
- The `android:host` attribute with a domain URI pattern
- The `category` element, which can be one of the following:
    - `android.intent.action.VIEW`
    - `android.intent.category.BROWSABLE`

The Digital Asset Links API

We can use the **Digital Asset Links API** to confirm that our link's JSON file is properly hosted and defined using the following syntax:

```
https://digitalassetlinks.googleapis.com/v1/statements:list?
  source.web.site=https://<DOMAIN>:<port>&
  relation=delegate_permission/common.handle_all_urls
```

Testing our intent

Now that we have confirmed that the hosted JSON file is valid, we will install the app on our device and wait for at least 20-30 seconds for the verification process to complete. After this, we can check whether the system has verified our app and set the correct link-handling policies using the following syntax:

```
adb shell am start -a android.intent.action.VIEW \
  -c android.intent.category.BROWSABLE \
  -d "http://<DOMAIN>:<port>"
```

For example, if we take YouTube videos, we can trigger the YouTube app to open the video directly using the following command:

```
adb shell am start -a android.intent.action.VIEW -c
android.intent.category.BROWSABLE -d "http://youtu.be/U9tw5ypqEN0"
```

Checking policies using adb

Android Debug Bridge (**adb**) can help us check existing link-handling policies for all applications in our device using the following command:

```
adb shell dumpsys package domain-preferred-apps
```

The following screenshot is the result of the preceding command:

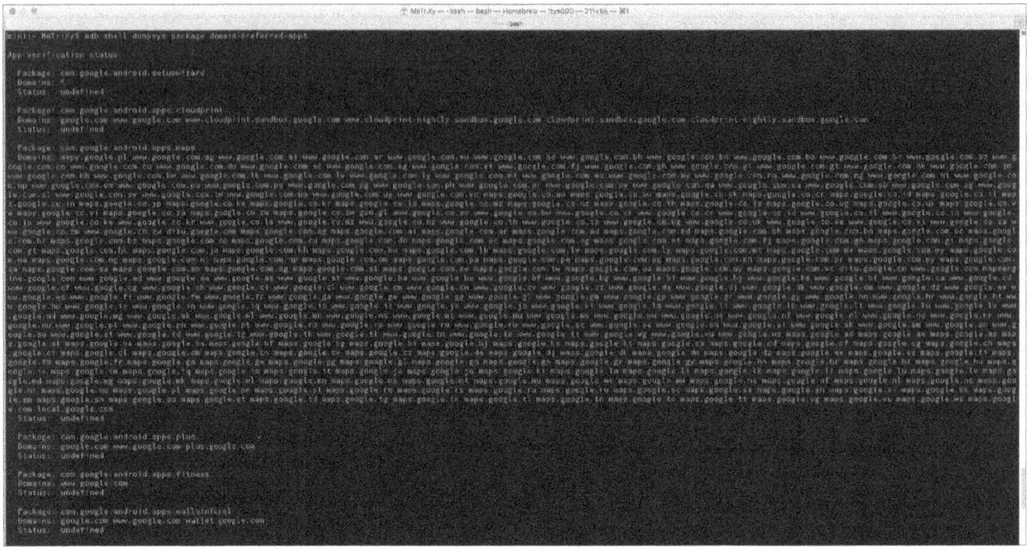

Another option is as follows:

```
adb shell dumpsys package d
```

The following screenshot is the result of the preceding command:

 We must wait at least 20-30 seconds after the installation for the system to complete the verification process.

The following listing indicates apps' association with domains per user:

- **Package**: This refers to the app's package name, as declared in its manifest
- **Domains**: This refers to the list of hosts whose web links are handled by this app; blank spaces are used as delimiters
- **Status**: This refers to the current link-handling setting for this app

Passing the verification and `android:autoVerify="true"` will show a status of `always`.

The hexadecimal number after the status (as shown in the preceding screenshot) is the Android system's record of the user's app linkage preferences. It does not indicate that the verification has succeeded.

The user can change the app link settings before the end of the verification process, which means that we may see a false positive for a successful verification. User preferences take precedence over programmatic verification, so we will see that the link goes directly to our app without displaying a dialog, as if verification had succeeded.

Summary

As we saw, app linking has become powerful in Android Marshmallow. This allows you, the app developers, help the system better decide how to act. Handling web URLs will give you wider exposure, a bigger funnel for your apps, and better experience, which you can provide to your users (which in turn leads to better ratings, more downloads, and vice versa).

App linking is simple to implement, easy to understand, and is a must-have feature in the mobile/web world today. While app linking enables better action handling for those using your applications, users can have multiple devices, expecting the same behavior on each device, and would be more engaged if their data and action handling happens smoothly. This brings us to our next chapter, which will teach you how to back up user settings and more.

3
Apps' Auto Backup

Have you ever taken the time to set up an app on your phone, use it for a while, pour in a lot of content, and switch phones due to a mishap just to discover that your data and settings have gone with the wind?

One of the key features of Android Marshmallow is that it supports full automatic data backup and restore for user apps. This improves the user experience, makes the overall engagement more fun, and shortens the boarding time for multiple devices. Like we discussed in the previous chapters, happy users lead to happy developers.

You can unload the burden of setting up a new device; it doesn't matter whether it's an added device or a replacement. The user will end up with the same app configuration and data, allowing work to be more **device-agnostic**. For this feature to be enabled on your applications, you must target the Android Marshmallow SDK's version 23; no extra code is needed by default even though you can configure the feature and allow specific behavior whenever required. Data is automatically restored when a user changes or upgrades the device.

In this chapter, we will learn how this feature works and configure the information that we want to back up. We'll cover the following topics:

- An overview
- Data backup configuration
- Backup configuration testing
- Important bytes

An overview

The automatic backup feature is created by taking the data created within your app and uploading it to the user's Google Drive account, keeping it encrypted. This doesn't affect the user's drive quota or your quota, for that matter. Each app is limited to 25 MB backup per user, and once you reach that amount, your app will stop backing up. Also, note that it's *completely free*!

Backup is done in cycles of 24 hours, nights only, and it's done automatically, usually when the device is idle, charging, and connected to a Wi-Fi network. These conditions are there for battery efficiency, data charges, and, of course, to keep the user interference to a minimum. Android systems have a **Backup Manager** service, which uploads all the available backup data to the cloud. Switching to a new device or uninstalling and reinstalling the app will trigger the restore operation, which in turn copies the data into the app's data directory.

 This new behavior allows you to keep using your existing backup service calls as well.

To read more about the **Android Backup Service** that was used prior to Android Marshmallow, head to:

```
https://developer.android.com/guide/topics/data/backup.html
```

Data backup configuration

We have a lot of data that we want to back up for our users, but we also don't want to back up all the data. Let's say we all agree not to back up users' passwords or other sensitive data, but what if you have a specific app configuration that is generated based on the device the user is using? This too should be excluded in a manner similar to device tokens such as **Google Cloud Messaging** (**GCM**) and others. I would recommend that you figure out which data your app keeps persistently and whether this data should and can be device-agnostic.

You can configure what is being backed up besides the automatically excluded files mentioned earlier. This configuration should be declared in your app's manifest via the `android:fullBackupContent` attribute. You will need to create a new XML file that should reside in your `res/xml` folder, and this will have specific rules for the backing up of your app's data.

Including or excluding data

XML file configuration includes a simple batch of `include`/`exclude` tags, which indicate whether or not you need to back up a directory or a specific file. Keep in mind that by default, the XML is *reductive*, which means that you back up everything possible unless there is an instruction to exclude it in your XML.

Another possible configuration is the constructive configuration in which you specify only the things you want to back up, and they will be added to the backup. This configuration behavior is done by adding an `include` tag to your XML, and from then onward, it will remain constructive.

As we can see in our example, we specify a backup scheme configuration in the app's manifest:

```xml
<?xml version="1.0" encoding="utf-8"?>
<manifest
  xmlns:android="http://schemas.android.com/apk/res/android"
  xmlns:tools="http://schemas.android.com/tools"
  package="com.yourapp.androidapp">
  <uses-sdk android:minSdkVersion="16" />
  <uses-sdk android:targetSdkVersion="23" />
  <app android:fullBackupContent="@xml/androidapp_backup_config">
  </app>
</manifest>
```

After declaring the file in our manifest, we also need to construct it in our `res/xml` folder; for example, take a look at the following:

```xml
<?xml version="1.0" encoding="utf-8"?>
<full-backup-content>
  <exclude domain="database" path="sensitive_database_name.db"/>
  <exclude domain="sharedpref"
    path="androidapp_shared_prefs_name"/>
  <exclude domain="file" path="some_file.file_Extension"/>
  <exclude domain="file" path="some_file.file_Extension"/>
</full-backup-content>
```

This example backup configuration excludes only specific data from being backed up. All other files are backed up.

The backup configuration syntax

Although you should've sorted out your app's specific persistent data, we can go over the configuration syntax that should be in the XML. The syntax for the configuration XML file is as follows:

```
<full-backup-content>
    <include domain=[ "root" | "sharedpref" | "database" | "file" |
        "external"] path="string" />
    <exclude domain=[ "root" | "sharedpref" | "database" | "file" |
        "external"] path="string" />
</full-backup-content>
```

Don't forget to read the explanation for each attribute and element here:

- `<include>`: You should use this tag whenever you want to specifically add a resource from any of the approved sorts to the backup. Remember that whenever you specify an `<include>` tag, the backup behavior changes to constructive, and the system only backs up resources specified with the `<include>` tags.

- `<exclude>`: You should use this tag whenever you want to exclude any of the app's resources from the backup. As mentioned earlier, you should exclude sensitive data and your app's device-specific data. Here, the behavior is like this: the system backs up all of your app's data except the resources specified with the `<exclude>` tag.

- `domain`: This appears on `include` as well as `exclude` tags. It allows you to declare the resource type you want to include or exclude from the backup. The domain has specific valid values that you can choose from:

    - `root`: This implies that the resource should be in the app's `root` directory

    - `file`: This implies that the resource is a file located in the `Files` directory and is accessible via the `getFilesDir()` method

    - `database`: This implies that your resource is a database file that can be located via the `getDatabasePath()` method or the `SQLiteOpenHelper` class

    - `sharedpref`: This implies that your resource is a `SharedPreferences` object that can be accessed via the `getSharedPreferences()` method

    - `external`: This implies that your resource is a file in an external storage located in a directory accessed by the `getExternalFilesDir()` method

    - `path`: This a `String` path to the resource that you want included in or excluded from backup

Opting out from app data backup

On some occasions, you might decide that you wish not to use the app data backup feature in your app. In such a situation, you will be able to notify the system that your app has opted out.

Setting the `android:allowBackup` attribute to `false` in your manifest is done using the following command:

```
android:allowBackup="false"
```

Backup configuration testing

By now, you have created a backup configuration and you might (should) test it and make sure that your app saves the data, restores it, and works without any issues.

Setting backup logs

Before you test your app's configuration, you might want to enable logging; this is done via `adb`, where you set the parser `log` property to VERBOSE:

```
$ adb shell setprop log.tag.BackupXmlParserLogging VERBOSE
```

Testing the backup feature can be split into two parts:

* Testing the backup phase
* Testing the restore phase

Testing the backup phase

The backup can be run manually, but first, you must run the Backup Manager via the `adb` command:

```
$ adb shell bmgr run
```

After the Backup Manager is up and running, we can trigger the backup phase via `adb` and run our app's package name as the `<PACKAGE.NAME>` parameter:

```
$ adb shell bmgr fullbackup <PACKAGE.NAME>
```

Testing the restore phase

We executed the backup phase and all went well. Now, we want to test the restore phase and verify that all the backed-up data is restored properly and we didn't miss out on any resource. We manually run a restore (*must* be done after your app data is backed up). This is done via the `adb` shell, specifying the package name for your app as the `<PACKAGE.NAME>` parameter:

```
$ adb shell bmgr restore <PACKAGE.NAME>
```

 The `restore` action stops your app and wipes its data before actually performing the restore operation.

Troubleshooting

Issues can occur in any place, including our case. If you run into issues, you should try and clear the data by turning backup on and off by navigating to **Settings | Backup & reset**, factory resetting the device:

You can clear the data using this command:

```
$ adb shell bmgr wipe <TRANSPORT> <PACKAGE.NAME>
```

The `<TRANSPORT>` tag is prefixed by `com.google.android.gms/`. To view the list of transports, you can run following `adb` command:

```
$ adb shell bmgr list transports
```

The following screenshot is the result of the preceding command:

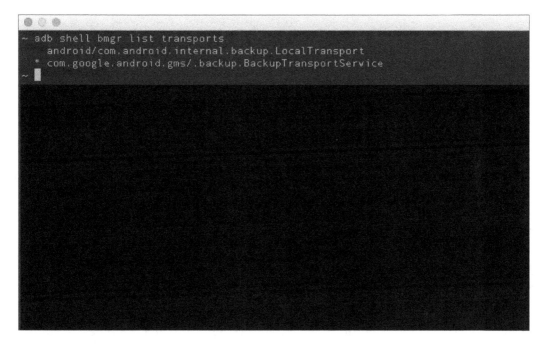

Important bytes

Before we jump into the next chapter, let's go through a couple of important subtopics within the Android apps' backup feature.

System backup does not include the following:

- Files located in `CacheDir` via the `getCacheDir()` method (API 1 and above)
- Files located in `CodeCacheDir` via the `getCodeCacheDir()` method (API 21 and above)

- Files located in the external storage and not in `ExternalFilesDir` via the `getExternalFilesDir(String type)` method, where the type can be as follows:
  - `null` for the root of the file directory
  - Any of these types for a specific subfolder/directory:
    - `android.os.Environment.DIRECTORY_MUSIC`
    - `android.os.Environment.DIRECTORY_PODCASTS`
    - `android.os.Environment.DIRECTORY_RINGTONES`
    - `android.os.Environment.DIRECTORY_ALARMS`
    - `android.os.Environment.DIRECTORY_NOTIFICATIONS`
    - `android.os.Environment.DIRECTORY_PICTURES`
    - `android.os.Environment.DIRECTORY_MOVIES`
- Files located in `NoBackupFilesDir` via the `getNoBackupFilesDir()` method (API 21 and above)

What to exclude from the backup

Though we have discussed this earlier, you may need to revise which app data is eligible for backup.

Among the excluded data, you must exclude any device-specific identifiers that are either issued by a server or generated on the device, including the GCM registration token.

You must also add the excluding logic for any account credentials or other sensitive information.

BackupAgent and backup events

You can implement your own `BackupAgent` attribute, which allows you to listen to events. `BackupAgent` has several callbacks that you can override, one of which is the `onRestoreFinished()` method, which is called after a successful restore takes place. You should add the `android:fullBackupOnly="true"` attribute to your manifest in addition to `android:backupAgent`; this will indicate that while your application has a `BackupAgent` attribute, Android Marshmallow and other devices will only perform full-data backup operations.

This technique can come in handy when you want to exclude a few keys from your **SharedPreferences** backup (device-specific tokens, GCM tokens, and so on). Instead of partitioning SharedPreferences into multiple files, you can simply remove the keys at restore time when `onRestoreFinished()` is called.

Keep in mind that other sensitive data is not supposed to be backed up anyway. You can read more about `BackupAgent` at:

`http://developer.android.com/reference/android/app/backup/` `BackupAgent.html`.

Summary

Android Marshmallow has brought in a great backup feature for apps, reducing friction for users migrating to new devices.

In a world full of diverse apps, maximizing the benefits from automatic backups leads to better user experience. The goal of this feature is to unload the burden and shorten the time required to set up a new device with the user's favorite apps.

Allowing the users to enter your app with merely a password prompt after the new installation can be a great experience; try it yourself! You can check out the sample code that's included or go to the GitHub repository at:

`https://github.com/MaTriXy/apps_autobackup_example`

In our next chapter, we will dive into more changes executed in Android Marshmallow as we unfold its awesomeness.

4
Changes Unfold

Android Marshmallow holds some changes that might get overlooked. A lot of these changes are short but will require your full attention to fully understand them and make sure you don't miss out when trying to use a removed/deprecated API, a new *flow*, or a new and improved API.

I've bundled up a group of changes that you might use or need to know and understand when building your applications for Android 6.0 (Marshmallow):

- Power-saving modes
- Removable storage adoption
- Apache HTTP client removal
- Notifications
- Text selection
- Support library notice
- Android Keystore changes
- Wi-Fi and networking changes
- Runtime
- Hardware identifier
- APK validation
- USB connection
- Direct Share
- Voice interactions
- The Assist API
- Bluetooth API changes

The preceding group doesn't include a separate chapter for major changes, for example, the permissions model covered in *Chapter 1*, *Android Marshmallow Permissions*, or an improved API, such as the video/audio/camera API that we will cover in the next chapter.

Power-saving modes

Android 6.0 has added new power-saving modes, **Doze** and **App Standby**, prolonged battery life by up to 2 times according to Google's measurements. The Doze mode has been created to improve the sleep efficiency of idle devices, while the App Standby mode has been designed to prevent apps from eating up power while in the idle state. On both occasions, plugging in the device to chargers allows normal operations to resume.

The Doze mode

Dozing is when a device is unplugged, the screen is off, and it's stationary (this can be determined via sensors, such as the accelerometer) for a determined period of time. What we get is a state where the system is kept in the sleep state as long as possible. When an Android 6.0 device is in the Doze mode, not much will happen in the background, as shown in the following figure:

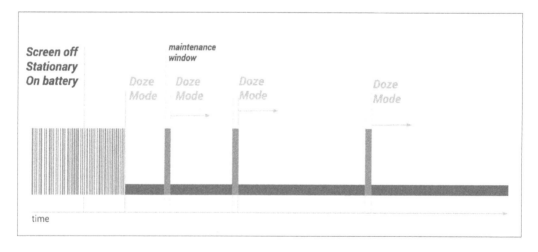

In short, everything you think will happen in the background will not actually happen.

What happens to apps when a device is dozing?

When a device enters the dozing state, you will encounter some battery-efficient system behavior, which will include the following:

- Network access is restricted unless your app receives a high-priority GCM
- **Wake locks** are ignored but are granted to apps
- Syncs and jobs are deferred using the following:
    - ° Sync adapters
    - ° JobScheduler (not allowed to run; this is enforced by the system)
- Alarms are deferred

 If you have important alarms and need to trigger the UI:

- Use the setAndAllowWhileIdle() method
- Can't be abused; this is allowed once every 15 minutes
- Wi-Fi scans are off
- GPS is off

The Doze mode will end shortly before any setAlarmClock() alarms; it can also end when the states of being stationary and unplugged are exchanged. Exiting the Doze mode will trigger the device to execute any jobs and syncs that are pending.

Testing apps with Doze mode

Test apps using your device (with Android 6.0) and adb commands:

1. Simulate an unplugged device using the following command:

   ```
   $ adb shell dumpsys battery unplug
   ```

   This will cause your battery icon to show as if the device is not plugged in.

2. Take the step to the next state using the following command:

```
$ adb shell dumpsys deviceidle step
```

This can be seen in the following screenshot:

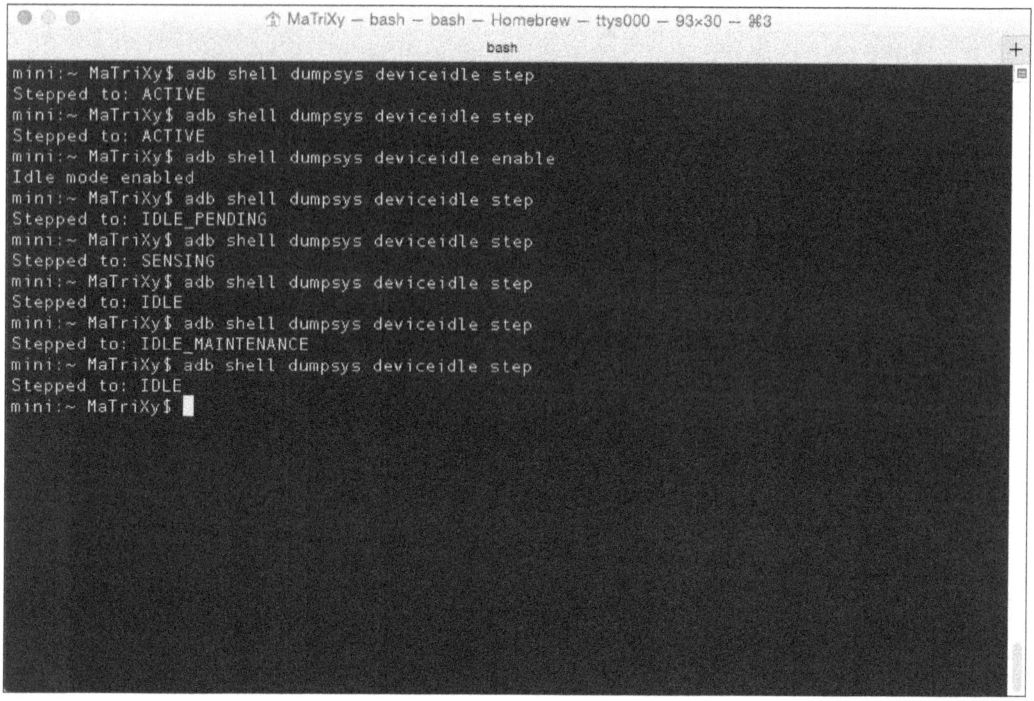

MaTriXy — bash — bash — Homebrew — ttys000 — 93×30 — ⌘3
bash
mini:~ MaTriXy$ adb shell dumpsys deviceidle step
Stepped to: ACTIVE
mini:~ MaTriXy$ adb shell dumpsys deviceidle step
Stepped to: ACTIVE
mini:~ MaTriXy$ adb shell dumpsys deviceidle enable
Idle mode enabled
mini:~ MaTriXy$ adb shell dumpsys deviceidle step
Stepped to: IDLE_PENDING
mini:~ MaTriXy$ adb shell dumpsys deviceidle step
Stepped to: SENSING
mini:~ MaTriXy$ adb shell dumpsys deviceidle step
Stepped to: IDLE
mini:~ MaTriXy$ adb shell dumpsys deviceidle step
Stepped to: IDLE_MAINTENANCE
mini:~ MaTriXy$ adb shell dumpsys deviceidle step
Stepped to: IDLE
mini:~ MaTriXy$

3. Reset the battery state back to its normal condition using the following command:

```
$ adb shell dumpsys battery reset
```

You can also list the available commands using the following command:

```
$ adb shell dumpsys deviceidle -h
```

This prints more information about the `deviceidle` usage, as shown in the following screenshot:

The App Standby mode

App Standby is a special mode that apps will be in when a system determines that an app is idle. An app is considered idle after a period of time unless the app exhibits the following features:

- It has a foreground process at that time (an activity or service)
- It displays notifications on the lock screen or in the notification tray
- It was explicitly launched by the user
- It was marked as excluded from optimizations via the settings app

What happens to apps when in the App Standby mode?

If the device is unplugged, syncs and jobs are deferred and network access is restricted.

If the device is plugged in, the system releases the app lock in the standby state, allowing the device to resume access to the network and/or execute any pending jobs and syncs.

 When in the idle state for a long period of time, the system allows idle apps to access the network just once a day.

Testing apps with the App Standby mode

Test apps using your device (with Android 6.0) and `adb` commands:

1. Simulate the app that's going into the standby mode:

    ```
    $ adb shell am broadcast -a android.os.action.DISCHARGING
    $ adb shell am set-inactive <App Package Name > true
    ```

2. Simulate by waking your app:

    ```
    $ adb shell am set-inactive <App Package Name > false
    ```

3. See what happens when your app awakens. Test recovering gracefully from standby mode. Check whether your app's notifications and background jobs function as you would anticipate.

You can set your app as inactive via the following command:

```
$ adb shell am set-inactive <App Package Name > true
```

You can also check the status of your app via the following command:

```
$ adb shell am get-inactive <App Package Name >
```

 The sample test was done on Google Photos behavior; all rights are reserved.

The console output, for example, is as follows:

```
~ adb shell am set-inactive com.google.android.apps.photos false
~ adb shell am get-inactive com.google.android.apps.photos
Idle=false
~ adb shell am set-inactive com.google.android.apps.photos true
~ adb shell am get-inactive com.google.android.apps.photos
Idle=true
```

Excluded apps and settings

You can exclude apps from the App Standby mode via the settings apps, as mentioned earlier. The procedure to do this is as follows:

1. Go to **Settings | Apps**.

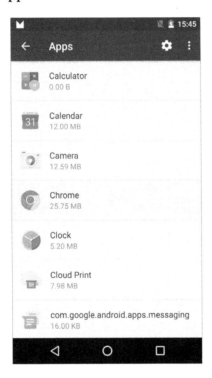

2. Click on the cog/gear icon to open the **Configure apps** screen.

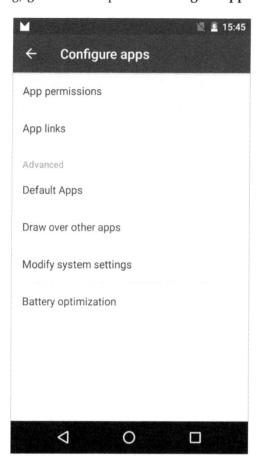

3. Choose **Battery optimization**.

4. The following screenshot shows a list of the apps excluded from the App Standby mode — that is, the ones that are not optimized. You can open the selection for all apps and choose the exact behavior you require for each application.

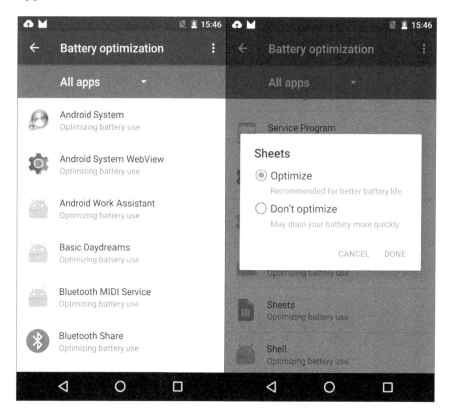

Tips

Here are a few points and tips for you to note and remember:

- Use `isIgnoringBatteryOptimizations()` with a `PowerManager` instance and check whether your app is on the **whitelist**

- Navigate the user directly to the configuration screen using the following:

```
startActivity(new
Intent(Settings.ACTION_IGNORE_BATTERY_OPTIMIZATION_SETTINGS
));
```

- Perform the following steps to display a system dialog asking about adding a specific app to the whitelist:

    1. Add the REQUEST_IGNORE_BATTERY_OPTIMIZATIONS permission to the application's manifest.

    2. Create a URI package pointing to your application.

    3. Wrap the URI in an intent and call startActivity() with it as shown in the following code:

    ```
    Intent intent = new
    Intent(Settings.ACTION_REQUEST_IGNORE_BATTERY_OPTIMIZATIONS
    , Uri.parse("package:" + getPackageName()));
    startActivity(intent);
    ```

- Note that if our app is already whitelisted, the dialog won't be displayed again

Removable storage adoption

Android Marshmallow allows users to *adopt* external storage devices, such as SD cards. Such adoptions will format and encrypt the storage device and mount it as internal storage. Once done, users can move apps and apps' private data between storage devices. The android:installLocation preference in the manifest will then be used by the system to determine the available locations for each app. What you need to keep in mind is that using Context methods for directories or files and ApplicationInfo fields will return values that can change between runs. You should always call these APIs dynamically. Don't use hardcoded file paths or persist fully qualified file paths.

The Context methods are as follows:

- getFilesDir()
- getCacheDir()
- getCodeCacheDir()
- getDatabasePath()
- getDir()
- getNoBackupFilesDir()
- getFileStreamPath()
- getPackageCodePath()
- getPackageResourcePath()

The `ApplicationInfo` fields are as follows:

- `dataDir`
- `sourceDir`
- `nativeLibraryDir`
- `publicSourceDir`
- `splitSourceDirs`
- `splitPublicSourceDirs`

You can debug this feature and enable the adoption of a USB drive connected via an **OTG** (short for **On The Go**) cable using the following command:

```
$ adb shell sm set-force-adoptable true
```

For more on USB, head to https://developer.android.com/guide/topics/connectivity/usb/index.html.

Apache HTTP client removal

The **Apache HTTP client** has been deprecated for quite some time — since 2011 or so. Using this client on Android 2.3 and higher was not recommended; now with Android 6.0 Marshmallow, this API has been removed. So, we'll use the `HttpURLConnection` class instead.

This API is more efficient, reduces network use, and minimizes power consumption.

If you wish to continue using the Apache HTTP APIs, you must first declare the following compile-time dependencies in your `build.gradle` file:

```
android {
  useLibrary 'org.apache.http.legacy'
}
```

If you have compile errors in the Android studio, you can head to these questions and solutions on stackoverflow:

- http://stackoverflow.com/q/30856785/529518
- http://stackoverflow.com/q/31653002/529518

Notifications

There are a few changes to the notifications feature, as follows:

- The `Notification.setLatestEventInfo()` method is now removed. When constructing notifications, we must use the `Notification.Builder` class.

- Updating a notification is also done via the `Notification.Builder` instance using the same instance of the builder, and calling the `build()` method will get us an updated `Notification` instance. If legacy support is required, you can use `NotificationCompat.Builder` instead, which is available in the **Android Support Library**.

- The `adb shell dumpsys notification` command no longer prints out notification text. The proper usage now is `adb shell dumpsys notification --noredact`.

- The newly added `INTERRUPTION_FILTER_ALARMS` filter level corresponds to a new mode: *Alarms only do not disturb*.

- The newly added `CATEGORY_REMINDER` category is used for user-scheduled reminders.

- The newly added `Icon` class allows icons to be attached to notifications via the `setSmallIcon()` and `setLargeIcon()` methods.

- The updated `addAction()` method now accepts an `Icon` object instead of a drawable resource ID.

- The newly added `getActiveNotifications()` method allows you to find out which notifications are currently alive.

- We can obtain some knowledge about what the user is and is not expecting to see under notifications when using the following methods:

  - The newly added `getCurrentInterruptionFilter()` method returns the current notification interruption filter in which notifications are allowed to interrupt the user

  - The newly added `getNotificationPolicy()` method returns the current notification policy

Text selection

Part of the material design guide specifications discuss text selection in your applications. Users select text within your app, and you now have an API to incorporate a **floating toolbar** design pattern that's similar to a contextual action bar. For more information about the design specifications, head to http://www.google.com/design/spec/patterns/selection.html#selection-item-selection.

The implementation steps are as follows:

1. Change your `ActionMode` calls to `startActionMode(Callback,ActionMode.TYPE_FLOATING)`.

2. Extend `ActionMode.Callback2`.

3. Override the `onGetContentRect()` method and provide coordinates for the content `Rect` object in the view.

4. Call the `invalidateContentRect()` method when you need to invalidate the `Rect` object and it's position is no longer valid.

Support library notice

Floating toolbars are not backward-compatible. `Appcompat` takes control over `ActionMode` objects by default. This will prevent floating toolbars from being displayed.

The implementation steps are as follows:

1. Call `getDelegate()` and `setHandleNativeActionModesEnabled()` on the returned `AppCompatDelegate` object.

2. Set the input parameter to `false`.

This call will return control of `ActionMode` objects to the framework, allowing 6.0 devices to support `ActionBar` or floating toolbar modes and allowing earlier versions to support the `ActionBar` modes.

Android Keystore changes

From Android 6.0 onward, the Android Keystore provider no longer supports **Digital Signature Algorithm (DSA)**.

For more information about keystore and its usage, visit `https://developer.android.com/training/articles/keystore.html`.

Wi-Fi and networking changes

Android Marshmallow has introduced a few changes to the Wi-Fi and networking APIs.

Changing the state of `WifiConfiguration` objects is only possible for self-created objects. You are restricted from modifying or deleting `WifiConfiguration` objects created by the user or other apps.

In earlier versions, forcing the device to connect to a specific Wi-Fi network using `enableNetwork()` and setting up `disableAllOthers=true` caused the device to disconnect from other networks. This does not happen in Android 6.0. With `targetSdkVersion <=20`, your app is pinned to use the selected Wi-Fi network. When `targetSdkVersion >=21`, you need to use the `MultiNetwork` APIs and ensure that network traffic is assigned to the proper network. For more information on the `MultiNetwork` API, refer to `https://developer.android.com/about/versions/android-5.0.html#Wireless`.

Runtime

The Android **ART** (short for **Android runtime**) runtime was also updated in Android Marshmallow, and the following are the updates:

- The `newInstance()` method: The **Dalvik** (another runtime) issue for the checking of access rules incorrectly was fixed. If you wish to override access checks, call the `setAccessible()` method with the input parameter set to `true`.

- Using the `v7 Appcompat` library or the `v7 Recyclerview` library? You must update to the latest version.

- Make sure that any custom classes referenced from XML are updated so that their class constructors are accessible.

- Behavior of the dynamic linker is updated.

- The ART runtime understands the difference between a library's `soname` and its path; search by `soname` is now implemented. There was an open bug with this issue that was fixed; if you wish to extend your reading, visit here:

  `https://code.google.com/p/android/issues/detail?id=6670`

Hardware identifier

Android 6.0 has introduced a major change for greater data protection; the `WifiInfo.getMacAddress()` and `BluetoothAdapter.getAddress()` methods now return a constant value of `02:00:00:00:00:00`, which means you can't rely on these methods to get information.

Now, when you're trying to use some of the methods in the API, you need to add permissions:

- `WifiManager.getScanResults()` and `BluetoothLeScanner.startScan()` need one of these two permissions granted:
  - ◦ The `ACCESS_FINE_LOCATION` permission
  - ◦ The `ACCESS_COARSE_LOCATION` permission

- `BluetoothDevice.ACTION_FOUND`: This must have the `ACCESS_COARSE_LOCATION` permission

 When a device running Android 6.0 (Marshmallow) initiates a background Wi-Fi or Bluetooth scan, external devices see the origin as a randomized MAC address.

APK validation

The platform now performs strict validation of **Android Package Kits** (**APKs**).

If a file declared in the manifest is not present in the APK itself, then the APK is considered corrupted. Removing contents from the APK requires re-signing of the APK.

USB connection

By default, the USB connection is charge-only. Users must now grant permissions to interact via the USB port. Your applications should take this into account and be aware that permissions might not be granted.

Direct Share

One of the best things about technology, in my humble opinion, is that it gives users great options to interact and benefit from them. **Direct Share** can be treated as a great addition to the list of merits, with great, fluid user experience all around the app world. So, what is Direct Share? Well, almost every app today uses some sort of information/data exchange with other applications on the user's device or with the outside world via the sharing mechanism. The sharing mechanism exposes a piece of information from one application to another. Usually, a user will interact with a few close companions (family, close friends, or colleagues), and this is where Direct Share comes to your aid.

Direct Share is about a set of APIs required to make sharing intuitive and quick for users. You define Direct Share targets that launch a specific activity in your app. These targets are shown in the **Share** menu, allowing faster sharing and fluid data flow.

With Direct Share, users can share content to targets — say, contacts in other apps.

The implementation steps are as follows:

1. Define a class that extends the `ChooserTargetService` class.

2. Declare your service in the manifest.

3. Specify the `BIND_CHOOSER_TARGET_SERVICE` permission and an intent filter `SERVICE_INTERFACE` action.

An example service declaration is as follows:

```
<service android:name=".MyChooserTargetService"
  android:label="@string/McTs_name" android:permission=
  "android.permission.BIND_CHOOSER_TARGET_SERVICE">
  <intent-filter>
    <action android:name=
      "android.service.chooser.ChooserTargetService"/>
  </intent-filter>
</service>
```

Now, we have a service declared and, for each target we want to expose, we add a `<meta-data>` element with the `android.service.chooser.chooser_target_service` name in your app manifest:

```
<activity android:name=".SampleDirectShareTarget"
  android:label="@string/SampleDirectShareTarget_name">
  <intent-filter>
    <action android:name="android.intent.action.SEND" />
  </intent-filter>
  <meta-data android:name=
    "android.service.chooser.chooser_target_service"
    android:value=".ChooserTargetService" />
</activity>
```

Let's take a look at the code in our service:

```
public class MyChooserTargetService extends ChooserTargetService {
  private String mDirectShareTargetName;
  private final int MAX_SHARE_TARGETS = 5;
```

```
@Override
public void onCreate() {
  super.onCreate();
  mDirectShareTargetName = "Sharing Person demo #%d";
}

@Override
public List < ChooserTarget > onGetChooserTargets(ComponentName
sendTarget, IntentFilter matchedFilter) {
  ArrayList < ChooserTarget > result = new
  ArrayList < ChooserTarget > ();
  for (int i = 1; i <= MAX_SHARE_TARGETS; i++) {
    result.add(buildTarget(i));
  }
  return (result);
}

private ChooserTarget buildTarget(int targetId) {
  String title = String.format(mDirectShareTargetName,
  targetId);
  Icon icon = Icon.createWithResource(this,
  R.drawable.share_target_picture);
  float target_value = ((float)(25 - targetId) / 25);
  ComponentName componentName = new
  ComponentName(MyChooserTargetService.this,
  TargetActivity.class);
  Bundle bundle = new Bundle();
  bundle.putInt("simple_key", targetId);
  return (new ChooserTarget(title, icon, target_value,
  componentName, bundle));
  }
}
```

You can head to the **gist** if you wish to better view the code; you can visit
`https://gist.github.com/MaTriXy/adeacdf5496bcdae5f42`.

You have to implement the `onGetChooserTargets()` method as it will be called
when direct-share is triggered. You return a list of `ChooserTarget` objects that
represent sharing entry points to your application. The `onGetChooserTargets()`
results are included along with the regular `ACTION_SEND` activity itself. So, we only
want `ChooserTarget` objects that improve the flow and not duplicates.

When creating several `ChooserTarget` objects, each of them will probably point to the same activity. You must ensure that the extras bundle will contain distinguishing information so that each request will be unique. *Do not* put custom `Parcelable` objects in this bundle as it will cause crashes. You can find out more about `ChooserTarget` at `https://developer.android.com/reference/android/service/chooser/ChooserTarget.html#ChooserTarget`.

What if we have nothing to share?

Sometimes, you won't have any direct-share targets for a particular request; then, returning an empty list would be great. You can also disable the service via `android:enabled="false"` if you know that no results will be available until future usage of the app. Another option is to enable the service just for Android 6.0. This can easily be done using Boolean resources:

- Let's add a Boolean resource named `is_share_targets_on`:
    - The default value is `res/values/bools.xml`; set it to `false`
    - Android 6.0 is API 23, so in `res/values-v23/bools.xml`, set it to `true`

- Add `android:enabled="@bool/is_share_targets_on"` to your service declaration

Direct Share best practices

The following are few of the best practices followed in Direct Share:

- Android 6.0 limits the number of share targets, only showing eight of them. Providing more than eight share targets will show the best eight according to the score.
- The `FAILED BINDER TRANSACTION` exception can pop in for a visit if your list of targets exceeds 1 MB.
- Try to limit/cap how many share targets you try to return from your `ChooserTargetService` class.
- Make sure your app's icon is shown properly as it will be applied as a badge over the icons that you use for Direct Share targets.

Voice interactions

Voice interactions usually originate from user voice action. However, the voice interaction activity starts without any user input. Android Marshmallow has a new voice interaction API that, together with **voice actions**, allows us to build conversational voice experiences into our apps. Use the `isVoiceInteraction()` method to determine whether an activity is triggered by a voice action. Then, you can use the `VoiceInteractor` class and interact with the user.

Don't get confused with the `isVoiceInteractionRoot()` method, which returns `true` only if the activity is also the root of a voice interaction. Here, you will get `true` if your activity was started directly by the voice interaction service and not by another activity (another app) while undergoing voice interaction.

A best practice would be to prompt the users and confirm that this is their intended action. You already know that voice input is invoked from **Google Now**, where you can open URLs with a simple voice input, such as `open android.com`. Now, you can invent new voice actions and register them with Google, driving traffic directly and specifically to your app.

To learn more about implementing voice actions, head to `https://developers. google.com/voice-actions/interaction/`.

The Assist API

Back in **Google I/O 2015**, we saw the *Now on Tap* feature, where Google Now could peek into a running app and provide contextual assistance. The `Assist` API offers a new way for users to engage through an **assistant**. The assistant must be enabled prior to using it, allowing it to be aware of the current context. Triggering the assistant is done by long-pressing the *Home* button, no matter which app is active.

You can opt out of this by setting the `WindowManager.LayoutParams.FLAG_SECURE` flag.

Opting in requires you to use the new `AssistContent` class.

In order for us to be able to feed additional context from our app to the assistant, we need to follow these steps:

1. Implement the `Application.OnProvideAssistDataListener` interface, which is called when the user requests assistance.
2. Register it using `Application.registerOnProvideAssistDataListener()`.

3. Override the `onProvideAssistData()` callback, which is called when the user requests assistance. It is used to build an `ACTION_ASSIST` intent with all of the context of the current app.

4. Override the `onProvideAssistContent()` callback; this is optional. It is called when the user requests assistance, allowing us to provide references to content related to the current activity.

5. When done, unregister yourself using `Application.unregisterOnProvideAssistDataListener()`.

Bluetooth API Changes

Besides the changes mentioned previously, Android Marshmallow 6.0 has introduced a few more changes to the Bluetooth API.

Bluetooth stylus support

Stylus has been here a for a while; Bluetooth stylus didn't have full support for specifications in versions before Android Marshmallow. You can pair and connect a compatible Bluetooth stylus with either a phone or a tablet. Because you are not bound just to touches on screen, you can fuse the position, pressure, and button state data, allowing more precise user input and experience. Your app can add a listener to the stylus buttons and act accordingly. Just use the `View.OnContextClickListener` and `GestureDetector.OnContextClickListener` objects in your activity.

In order to detect stylus button interactions and movement, you need the following:

- The `MotionEvent` methods
- The `getTooltype()` method, which returns `TOOL_TYPE_STYLUS` if a stylus with a button is touched on the screen
- The `getButtonState()` method, which returns (on Android 6.0-targeted apps) the following:
    - `BUTTON_STYLUS_PRIMARY`: Press the primary stylus button
    - `BUTTON_STYLUS_SECONDARY`: Press the secondary button
    - `BUTTON_STYLUS_PRIMARY | BUTTON_STYLUS_SECONDARY`: Press both the buttons

- Targeted apps with a lower API level than Android 6.0 will result in the following:
  - ○ BUTTON_SECONDARY: Press the primary stylus button
  - ○ BUTTON_TERTIARY: Press the secondary button
  - ○ BUTTON_SECONDARY | BUTTON_TERTIARY: Press both the buttons

Improved Bluetooth low energy scanning

Used to Bluetooth low energy in your app? Well, now the scanning process is easier and improved. Use the new setCallbackType() method and specify that you want a callback when the system finds/sees an advertisement packet matching the ScanFilter class. You get more power-efficiency than in previous Android versions.

Summary

We went over a few of the changes in Android Marshmallow. All of these changes are important to follow and will help you in your app development cycles. There are a few more changes to be discussed in future chapters in a more detailed manner. Our next chapter talks about audio, video, and camera features and the changes made in Android 6.0.6.

5
Audio, Video, and Camera Features

Android Marshmallow gives us good audio, video, and camera capabilities, and you can see that improvements have been made to enable and better support new or mint condition protocols or even change the behavior of some APIs, such as the camera service.

In this chapter, we will try and explain these changes with a proper discussion on their usage and benefits. Our journey in the upcoming pages will cover the following topics:

- Audio features
- Video features
- Camera features

Audio features

Android Marshmallow 6.0 adds some enrichments to the audio features that we will cover in the upcoming sections.

Support for the MIDI protocol

The `android.media.midi` package was added in Android 6.0 (API 23).

With the new midi APIs, you can now send and receive **MIDI** (short for **Musical Instrument Digital Interface**) events in a much simpler way than earlier.

The package was built to provide us with capabilities to do the following:

- Connect and use a MIDI keyboard
- Connect to other MIDI controllers
- Use external MIDI synthesizers, external peripherals, lights, show control, and so on
- Allow dynamic music generation from games or music-creation apps
- Allow the creation and passing of MIDI messages between apps
- Allow Android devices to act as multi-touch controllers when connected to a laptop

When dealing with MIDI, you must declare it in the manifest, as follows:

```
<uses-feature android:name="android.software.midi"
  android:required="true"/>
```

Pay attention to the `required` part; in a manner similar to other features, setting it to `true` will make your app visible in the play store only if the device supports the MIDI API.

You can also check in runtime for MIDI support and then change the required part to `false`:

```
PackageManager pkgMgr = context.getPackageManager();
if (pkgMgr.hasSystemFeature(PackageManager.FEATURE_MIDI)) {
  //we can use MIDI API here as we know the device supports the
    MIDI API.
}
```

MidiManager

A way to properly use the MIDI API is via the `MidiManager` class; obtain it via `context` and use it when required:

```
MidiManager midiMgr =
  (MidiManager)context.getSystemService(Context.MIDI_SERVICE);
```

For more information, you can refer to:

`https://developer.android.com/reference/android/media/midi/package-summary.html`

Digital audio capture and playback

Two new classes have been added for digital audio capture and playback:

- `android.media.AudioRecord.Builder` - digital audio capture
- `android.media.AudioTrack.Builder` - digital audio playback

These will help configure the audio source and sink properties.

Audio and input devices

The new `hasMicrophone()` method has been added to the `InputDevice` class. This will report whether the device has a built-in microphone that developers can use. Let's say you want to enable voice search from a controller connected to Android TV and you get an `onSearchRequested()` callback for the user's search. You can then verify that there's a microphone with the `inputDevice` object you get in the callback.

Information on audio devices

The new `AudioManager.getDevices(int flags)` method allows easy retrieval of all the audio devices currently connected to the system. If you want to be notified when there are audio device connections/disconnections, you can register your app to an `AudioDeviceCallback` callback via the `AudioManager.registerAudioDevice Callback(AudioDeviceCallback callback, Handler handler)` method.

Changes in AudioManager

Some changes have been introduced in the `AudioManager` class, and they are as follows:

- Using `AudioManager` to set the volume directly is not supported.
- Using `AudioManager` to mute specific streams is not supported.
- The `AudioManager.setStreamSolo(int streamType, boolean state)` method is deprecated. If you need exclusive audio playback, use `AudioManager.requestAudioFocus(AudioManager. OnAudioFocusChangeListener l, int streamType, int durationHint)`.
- The `AudioManager.setStreamMute(int streamType, boolean state)` method is deprecated. If you need to use `AudioManager. adjustStreamVolume(int streamType, int direction, int flags)` for direction, you can use one of the newly added constants.

- `ADJUST_MUTE` will mute the volume. Note that it has no effect if the stream is already muted.

- `ADJUST_UNMUTE` will unmute the volume. Note that it has no effect if the stream is not muted.

Video features

In Android Marshmallow, the video processing API has been upgraded with new capabilities. Some new methods and even a new class has been added just for developers.

android.media.MediaSync

The all new `MediaSync` class has been designed to help us with synchronous audio and video streams' rendering. You can also use it to play audio- or video-only streams. You can use the dynamic playback rate and feed the buffers in a nonblocking action with a callback return. For more information on the proper usage, read:

`https://developer.android.com/reference/android/media/MediaSync.html`

MediaCodecInfo.CodecCapabilities. getMaxSupportedInstances

Now, we have a `MediaCodecInfo.CodecCapabilities.getMaxSupportedInstances` helper method to get the maximum number of supported concurrent codec instances. However, we must consider this only an upper bound. The actual number of concurrent instances can be lower depending on the device and the amount of available resources at the time of usage.

Why do we need to know this?

Let's think of a case where we have a media-playing application and we want to add effects between the movies played. We will need to use more than one video codec, decode two videos, and encode one video stream back to be displayed on screen. Checking with this API will allow you to add more features that rely upon multiple instances of codecs.

MediaPlayer.setPlaybackParams

We can now set the media playback rate for fast or slow motion playback. This will give us the chance to create a funny video app where we slow down parts or play them fast, creating a new video while playing. Audio playing is synced accordingly, so you might hear a person talking slowly or even fast, for that matter.

Camera features

In Android Lollipop, there was the new `Camera2` API, and now, in Android Marshmallow, there are a few more updates to the camera, flashlight, and image reprocessing features.

The flashlight API

Almost every device today has a camera, and almost every camera device has a flash unit. The `setTorchMode()` method has been added to control the flash torch mode.

The `setTorchMode()` method is used in the following manner:

```
CameraManager.setTorchMode (String cameraId, boolean enabled)
```

The `cameraId` element is the unique ID for the flash unit camera with which you want to change the torch mode. You can use `getCameraIdList()` to get the list of cameras and then use `getCameraCharacteristics(String cameraId)` to check whether flash is supported in that camera. The `setTorchMode()` method allows you to turn it on or off without opening the camera device and without requesting permission from the camera. The torch mode will be switched off as soon as the camera device becomes unavailable or when other camera resources that have the torch on become unavailable. Other apps can use the flash unit as well, so you need to check the mode when required or register a callback via the `registerTorchCallback()` method.

Refer to the sample app, **Torchi**, to see the entire code at:

`https://github.com/MaTriXy/Torchi`

 Turning on the torch mode may fail if the camera or other camera resources are in use.

The reprocessing API

As mentioned earlier, the `Camera2` API was given a few boosts to allow added support for **YUV** and private opaque format image reprocessing. Before using this API, we need to check whether these capabilities are available. This is why we use the `getCameraCharacteristics(String cameraId)` method and check for the `REPROCESS_MAX_CAPTURE_STALL` key.

android.media.ImageWriter

This is a new class that's been added to Android 6.0.

It allows us to create an image and feed it into a surface and then back to `CameraDevice`. Usually, `ImageWriter` is used along with `ImageReader`.

android.media.ImageReader

This is a new class that's been added to Android 6.0.

It allows us direct access to the image data rendered in a surface. `ImageReader`, along with `ImageWriter`, allows our app to create an image feed from the camera to the surface and back to the camera for reprocessing.

Changes in the camera service

Android Marshmallow has made a change to the *first come, first serve* access model; now, the service access model has favorites processes — ones that are marked as high-priority. This change results in some more logic-related work for us developers. We need to make sure that we take into account a situation where we get bumped up (higher priority) or debunked (lower priority due to a change in our application).

Let's try and explain this in a few simple bullets:

- When you want to access camera resources or open and configure a camera device, your access is verified according to the *priority* of your application process. An application process with foreground activities (visible user) is normally given a higher priority, which in turn allows a better chance to get the desired access when needed.

- On the other side of the *priority* scale, you can find low-priority apps that can and will be tossed aside (revoked from access) when a high-priority application attempts to use the camera. For example, when using the `Camera` API, you will get the `onError()` call when evicted, and when using the `Camera2` API, you will get the `onDisconnected()` call when evicted.

- Some devices out in the wild can allow separate applications to open and use separate camera devices simultaneously. The camera service now detects and disallows performance issues that are caused due to multiprocess usage. When the service detects such an issue, it will evict low-priority apps even if only one app is using that camera device.

- In a multiuser environment, when switching users, all active apps using the camera in the previous user profile will be evicted in order to allow proper usage and access to apps for the current user. This means that switching users will stop the camera-using apps from using the camera for sure.

Summary

In this chapter, we covered quite a few changes in and additions to the Android APIs. Android Marshmallow is more about helping us, the developers, achieve better media support and showcase our ideas when using the audio, video, or camera APIs.

In the next chapter, we will go over some of the Android features to understand the features, additions, and changes made.

6
Android for Work

Most of you know that Android devices have a huge market share percentage worldwide, and more and more businesses are following the **BYOD** (short for **Bring Your Own Device**) policy. This is done with the help of **Android for Work**, a special program for companies where several added features in the Android platform allow better mobile device management, administration, and integration within the company.

When dealing with enterprises or even small-sized and medium-sized businesses, you need to follow specific guidelines and harness the Android API to your benefit. You can read more about Android for Work at:

```
http://developer.android.com/training/enterprise/index.html
```

Android Marshmallow has made a few changes to the Android for Work program, where a lot of the changes were made for better and easier usage for developers as well as work users.

In this chapter, we will cover the Android Marshmallow changes that reflect or are related directly to Android for Work:

- Behavioral changes
- Single-use device improvements
- Silently installing/uninstalling apps
- Improved certificate access
- Automatic system updates
- Third-party certificate installation
- Data usage statistics
- Managing runtime permissions
- VPN access and display
- Work profile status

Behavioral changes

Android Marshmallow has introduced a few behavioral changes related to Android for Work.

The work profile contacts display option

Using the following setting, you can now display your work profile contacts in the dialer call log:

```
DevicePolicyManager.setCrossProfileCallerIdDisabled(ComponentName
    admin, boolean disabled)
```

You can also display the work contacts over Bluetooth with the new option. Setting this to `false` will allow the display; the default value is `true` (disabling the contact-sharing option):

```
DevicePolicyManager.setBluetoothContactSharingDisabled(ComponentNa
    me admin, boolean disabled)
```

Wi-Fi configuration options

When adding a Wi-Fi network via a work profile, usually, added configurations stay persistent even after the profile is deleted. Now, all configurations added by a profile owner are removed if the work profile is deleted.

The Wi-Fi configuration lock

A new `Settings.Global` setting has been added:

```
WIFI_DEVICE_OWNER_CONFIGS_LOCKDOWN
```

This setting is an integer value setting, which means that a zero value or absence will lead to all Wi-Fi configurations being modified or deleted by the user. Setting the integer value to a nonzero value will initiate the lock, which means that the user can't modify or delete Wi-Fi configurations created by a device owner — user-created configurations will still be modifiable. Note that an active device owner has complete privileges in any Wi-Fi configurations, even those not created by them.

Work Policy Controller addition

You can continue to add Google accounts to the device, but now, when adding an account that is managed by **Work Policy Controller**, the flow is changed to include the Work Policy Controller addition. An added account prompts the user to install the appropriate Work Policy Controller. This is also true when adding an account through settings or via the start up device's setup wizard. For more information on how to build a Work Policy Controller, read:

`http://developer.android.com/training/enterprise/work-policy-ctrl.html`

DevicePolicyManager changes

In `DevicePolicyManager`, you may encounter quite a few changes in behavior; these are listed in the following bullets with a short explanation:

- `setCameraDisabled()` affects the camera just for the calling user; if the profile is a managed profile, then the call doesn't affect the camera apps running on the primary user.

- `setKeyguardDisabledFeatures()` was made available for profile owners and device owners.

- Profile owners can set keyguard restrictions via the following:
    - `KEYGUARD_DISABLE_TRUST_AGENTS`: This will ignore the trust agent state on the keyguard on secure screens (the PIN code, pattern, or the password screen)
    - `KEYGUARD_DISABLE_FINGERPRINT`: This will disable the fingerprint sensor on the keyguard on secure screens (PIN code, pattern, or the password screen)
    - `KEYGUARD_DISABLE_UNREDACTED_NOTIFICATIONS`: This will allow only redacted notifications on secure keyguard screens and only notifications generated by applications in the managed profile

- `createAndInitializeUser()` is deprecated now.

- `createUser()` is deprecated now.

- Using the `setScreenCaptureDisabled()` method, the `Assist` feature is blocked, but this happens only when an app of the given user is in the foreground.

- `EXTRA_PROVISIONING_DEVICE_ADMIN_PACKAGE_CHECKSUM` is SHA-256 now. Legacy support for SHA-1 still exists, but it will be removed in future versions according to the documentation.

- `EXTRA_PROVISIONING_DEVICE_ADMIN_SIGNATURE_CHECKSUM` is SHA-256 only now.

- `EXTRA_PROVISIONING_RESET_PROTECTION_PARAMETERS` was removed so that NFC bump provisioning would not unlock a factory-reset-protected device.

- Passing data to the device owner during NFC provisioning can be done with `EXTRA_PROVISIONING_ADMIN_EXTRAS_BUNDLE`.

- New `DevicePolicyManager` API for permissions under Android Marshmallow's new permission model. You can read more about `DevicePolicyManager` at `https://developer.android.com/reference/android/app/admin/DevicePolicyManager.html`.

- `RESULT_CANCELED` is now returned if users cancel the setup flow initiated through an `ACTION_PROVISION_MANAGED_PROFILE` or `ACTION_PROVISION_MANAGED_DEVICE` intent.

- Changes to `Settings.Global`.

- Disabled the following set of settings via `setGlobalSettings()`:
  - `BLUETOOTH_ON`
  - `DEVELOPMENT_SETTINGS_ENABLED`
  - `MODE_RINGER`
  - `NETWORK_PREFERENCE`
  - `WIFI_ON`

- Enabled the `WIFI_DEVICE_OWNER_CONFIGS_LOCKDOWN` setting via `setGlobalSettings()`.

Single-use device improvements

You as the device owner can now control added settings, thus improving device management using the following:

- `setKeyguardDisabled()` can be used to disable or re-enable the keyguard

- `setStatusBarDisabled()` can be used to disable or re-enable the status bar

- `UserManager.DISALLOW_SAFE_BOOT` is a new constant that states whether the user can boot a device to safe boot

- `Settings.Global.STAY_ON_WHILE_PLUGGED_IN` will prevent the screen from turning off while plugged in to power

Silently installing/uninstalling apps

Now, you can silently install and uninstall applications using `PackageInstaller` APIs. This means installing apps without user interaction or even removing apps as part of the company policy. This feature enables you to use devices without actually activating a Google account. **Google Play for Work** is not required, allowing you to use devices as **kiosks**, showcasing specific apps not released yet, and so on.

Improved certificate access

Allowing users to grant managed apps' access to certificates without user interaction was not possible prior to Android Marshmallow, so now, a new callback has been added:

```
DeviceAdminReceiver.onChoosePrivateKeyAlias (Context context,
    Intent intent, int uid, Uri uri, String alias)
```

This callback will allow the device owner to provide the alias silently to the requesting application.

Automatic system updates

The following option has been added in Android 6.0 and its main purpose is to allow device owners to auto-accept a system update:

```
DevicePolicyManager.setSystemUpdatePolicy (ComponentName admin,
    SystemUpdatePolicy policy)
```

`SystemUpdatePolicy` has been added as well, and you can choose from three options:

- `TYPE_INSTALL_AUTOMATIC`: update as soon as you get an update
- `TYPE_INSTALL_WINDOWED`: update should be done within a timed system maintenance and only then, just for 30 days and then return to normal behavior
- `TYPE_POSTPONE`: postpone updates for up to 30 days and then return to normal behavior afterwards

This can come in handy if you have devices such as showcase tablets or kiosk mode devices, where the update should not mess with the devices' work.

Third-party certificate installation

Third-party apps now have the ability to call `DevicePolicyManager` APIs:

- `getInstalledCaCerts()`
- `hasCaCertInstalled()`
- `installCaCert()`
- `uninstallCaCert()`
- `uninstallAllUserCaCerts()`
- `installKeyPair()`

These API calls can only be done if the permission has been granted by the device owner or profile owner.

Data usage statistics

A new class has been added in Android 6.0: `NetworkStatsManager`. This helps you query for data usage statistics that can be seen in **Settings | Data usage**.

Access for profile owners is automatically granted in order for them to query data on their profile. Device owners get access to the data usage of the managed primary user.

 The `android.app.usage.NetworkUsageStats` class has been renamed to `NetworkStats`.

Managing runtime permissions

Android Marshmallow introduced the runtime permissions model, and Android for Work had to deal with managing policies for devices. You as device owner can now set a policy for all runtime requests of all applications using `setPermissionPolicy()`.

You can choose to prompt users to grant permissions or automatically grant or deny the permissions silently. The automatic policy means that the user cannot modify the app's permissions screen in **Settings**.

VPN access and display

When heading to **Settings** | **More** | **VPN**, you can now view the VPN apps. When using VPN, the notifications shown are now specific to how that VPN is configured:

- **The profile owner**: Notifications are shown according to the VPN configuration and based on the profile (personal, work, or both)
- **The device owner**: Notifications are shown when the VPN is configured for the entire device

Work profile status

Two new additions were introduced for the users to know that they are under a different profile:

- When using an app from a work profile, the status bar will display a briefcase icon
- When unlocking a device straight from a work profile app, a popup is displayed, alerting the user that this app runs on the work profile

Summary

As we saw in this chapter, Android Marshmallow has brought in quite a few changes to the Android for Work world. As developers, we need to always maintain a viable connection to the needs of an organization. We need to make sure we go over and understand the Android for Work world; the changes in Marshmallow help us build and target enterprise workflows with the benefit of a simpler API.

In the next chapter, we will learn about the Chrome custom tabs API's usage and flow.

7
Chrome Custom Tabs

Have you ever wanted to add a **WebView** to your application? Maybe you've wanted to add browsing for a few web pages and show some relevant content from within your application? I know I had to. On almost every occasion, I was reluctant to use the WebView feature as this was one of the ugliest parts of the app.

You can clearly see that the WebView feature is a web portion and the UI was added quite a few Android versions back, which caused my OCD UI/UX sense go kaboom. One of the newest additions released by Google was **Chrome custom tabs**.

In this chapter, we will explore Chrome custom tabs and try to explain and demonstrate the benefits of using it instead of the plain old WebView:

- What is a Chrome custom tab?
- When to use Chrome custom tabs
- The implementation guide

What is a Chrome custom tab?

Well, most of us know tabs from every day Internet browsing. It doesn't really matter which browser you use; all browsers support tabs and multiple tabs' browsing. This allows us to have more than one website open at the same time and navigate between the opened instances. In Android, things are much the same, but when using WebView, you don't have tabs.

What is WebView?

WebView is the part in the Android OS that's responsible for rendering web pages in most Android apps. If you see web content in an Android app, chances are you're looking at WebView. The major exceptions to this rule are some of the Android browsers, such as Chrome, Firefox, and so on.

In Android 4.3 and lower, WebView uses code based on Apple's **Webkit**. In Android 4.4 and higher, WebView is based on the **Chromium** project, which is the open source base of Google Chrome. In Android 5.0, WebView was decoupled into a separate app that allowed timely updates through Google Play without requiring firmware updates to be issued, and the same technique was used with Google Play services.

Now, let's talk again about a simple scenario: we want to display web content (URL-related) in our application. We have two options: either launch a browser or build our own in-app browser using WebView. Both options have trade-offs or disadvantages if we write them down. A browser is an external application and you can't really change its UI; while using it, you push the users to other apps and you may lose them in *the wild*. On the other hand, using WebView will keep the users tightly inside. However, actually dealing with all possible actions in WebView is quite an overhead.

Google heard our rant and came to the rescue with Chrome custom tabs. Now we have better control over the web content in our application, and we can stitch web content into our app in a cleaner, prettier manner.

Customization options

Chrome custom tabs allow several modifications and tweaks:

- The toolbar color
- Enter and exit animations
- Custom actions for the toolbar and overflow menu
- Prestarted and prefetched content for faster loading

When to use Chrome custom tabs

Ever since WebView came out, applications have been using it in multiple ways, embedding content—local static content inside the APK and dynamic content as loading web pages that were not designed for mobile devices at the beginning. Later on we saw the rise of the mobile web era complete with hybrid applications).

Chrome custom tabs are a bit more than just loading local content or mobile-compatible web content. They should be used when you load web data and want to allow simple implementation and easier code maintenance and, furthermore, make the web content part of your application—as if it's always there within your app.

Among the reasons why you should use custom tabs are the following:

- Easy implementation: you use the support library when required or just add extras to your `View` intent. It's that simple.

- In app UI modifications, you can do the following:
    - Set the toolbar color
    - Add/change the action button
    - Add custom menu items to the overflow menu
    - Set and create custom in/out animations when entering the tab or exiting to the previous location

- Easier navigation and navigation logic: you can get a callback notifying you about an external navigation, if required. You know when the user navigates to web content and where they should return when done.

- Chrome custom tabs allow added performance optimizations that you can use:
    - You can keep the engine running, so to speak, and actually give the custom tab a head start to start itself and do some warm up prior to using it. This is done without interfering or taking away precious application resources.
    - You can provide a URL to load in advance in the background while waiting for other user interactions. This speeds up the user-visible page loading time and gives the user a sense of blazing fast application where all the content is just a click away.

- While using the custom tab, the application won't be evicted as the application level will still be in the foreground even though the tab is on top of it. So, we remain at the top level for the entire usage time (unless a phone call or some other user interaction leads to a change).

- Using the same Chrome container means that users are already signed in to sites they connected to in the past; specific permissions that were granted previously apply here as well; even fill data, autocomplete, and sync work here.

- Chrome custom tabs allow us give the users the latest browser implementation on pre-Lollipop devices where WebView is not the latest version.

The implementation guide

As discussed earlier, we have a couple of features integrated into Chrome custom tabs. The first customizes the UI and interaction with the custom tabs. The second allows pages to be loaded faster and keeps the application alive.

Can we use Chrome custom tabs?

Before we start using custom tabs, we want to make sure they're supported. Chrome custom tabs expose a service, so the best check for support is to try and bind to the service. Success means that custom tabs are supported and can be used. You can check out this gist, which shows a helper how to to check it, or check the project source code later on at:

```
https://gist.github.com/MaTriXy/5775cb0ff98216b2a99d
```

After checking and learning that support exists, we will start with the UI and interaction part.

Custom UI and tab interaction

Here, we will use the well-known `ACTION_VIEW` intent action, and by appending extras to the intent sent to Chrome, we will trigger changes in the UI. Remember that the `ACTION_VIEW` intent is compatible with all browsers, including Chrome. There are some phones without Chrome out there, or there are instances where the device's default browser isn't Chrome. In these cases, the user will navigate to the specific browser application.

Intent is a convenient way to pass that extra data we want Chrome to get.

Don't use any of these flags when calling to the Chrome custom tabs:

- `FLAG_ACTIVITY_NEW_TASK`
- `FLAG_ACTIVITY_NEW_DOCUMENT`

Before using the API, we need to add it to our `gradle` file:

```
compile 'com.android.support:customtabs:23.1.0'
```

This will allow us to use the custom tab support library in our application:

```
CustomTabsIntent.EXTRA_SESSION
```

The preceding code is an extra from the custom tabs support library; it's used to match the session. It must be included in the intent when opening a custom tab. It can be null if there is no need to match any service-side sessions with the intent.

We have a sample project to show the options for the UI called
ChubbyTabby at `https://github.com/MaTriXy/ChubbyTabby`.

We will go over the important parts here as well. Our main interaction comes from
a special builder from the support library called `CustomTabsIntent.Builder`; this
class will help us build the intent we need for the custom tab:

```
CustomTabsIntent.Builder intentBuilder = new
  CustomTabsIntent.Builder(); //init our Builder

//Setting Toolbar Color
int color = getResources().getColor(R.color.primary);

//we use primary color for our toolbar as well - you can define any
color you want and use it.
intentBuilder.setToolbarColor(color);

//Enabling Title showing
intentBuilder.setShowTitle(true);

//this will show the title in the custom tab along the url showing at
the bottom part of the tab toolbar.

//This part is adding custom actions to the over flow menu
String menuItemTitle = getString(R.string.menu_title_share);
PendingIntent menuItemPendingIntent = createPendingShareIntent();
intentBuilder.addMenuItem(menuItemTitle, menuItemPendingIntent);
String menuItemEmailTitle = getString(R.string.menu_title_email);
PendingIntent menuItemPendingIntentTwo =
  createPendingEmailIntent();
intentBuilder.addMenuItem(menuItemEmailTitle,
  menuItemPendingIntentTwo);

//Setting custom Close Icon.
intentBuilder.setCloseButtonIcon(mCloseButtonBitmap);

//Adding custom icon with custom action for the share action.
intentBuilder.setActionButton(mActionButtonBitmap,
  getString(R.string.menu_title_share),
  createPendingShareIntent());

//Setting start and exit animation for the custom tab.
intentBuilder.setStartAnimations(this, R.anim.slide_in_right,
  R.anim.slide_out_left);
```

```
intentBuilder.setExitAnimations(this,
    android.R.anim.slide_in_left, android.R.anim.slide_out_right);
CustomTabActivityHelper.openCustomTab(this, intentBuilder.build(),
    Uri.parse(URL), new WebviewFallback(), useCustom);
```

A few things to notice here are as follows:

- Every menu item uses a pending intent; if you don't know what a pending intent is, head to:

  `http://developer.android.com/reference/android/app/`
  `PendingIntent.html`

- When we set custom icons, such as *close* buttons or an *action* button, for that matter, we use bitmaps and we must decode the bitmap prior to passing it to the builder

- Setting animations is easy and you can use animations' XML files that you created previously; just make sure that you test the result before releasing the app

The following screenshot is an example of a Chrome custom tab:

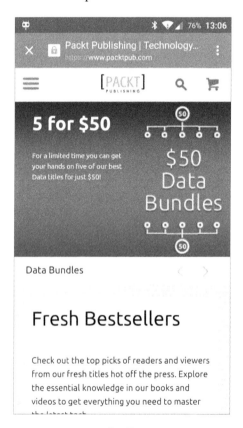

The custom action button

As developers, we have full control over the action buttons presented in our custom tab. For most use cases, we can think of a share action or maybe a more common option that your users will perform. The action button is basically a bundle with an icon of the action button and a pending intent that will be called by Chrome when your user hits the action button. The icon should be 24 dp in height and 24-48 dp in width according to specifications.

```
//Adding custom icon with custom action for the share action
intentBuilder.setActionButton(mActionButtonBitmap,
  getString(R.string.menu_title_share),
  createPendingShareIntent());
```

Configuring a custom menu

By default, Chrome custom tabs usually have a three-icon row with **Forward**, **Page Info**, and **Refresh** on top at all times and **Find in page** and **Open in Browser** (**Open in Chrome** can appear as well) at the footer of the menu.

We, developers, have the ability to add and customize up to three menu items that will appear between the icon row and foot items as shown in the following screenshot:

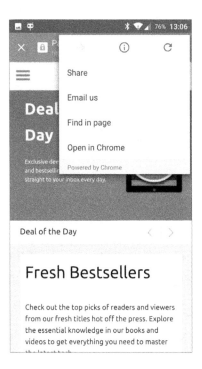

The menu we see is actually represented by an array of bundles, each with menu text and a pending intent that Chrome will call on our behalf when the user taps the item:

```
//This part is adding custom buttons to the over flow menu
String menuItemTitle = getString(R.string.menu_title_share);
PendingIntent menuItemPendingIntent = createPendingShareIntent();
intentBuilder.addMenuItem(menuItemTitle, menuItemPendingIntent);
String menuItemEmailTitle = getString(R.string.menu_title_email);
PendingIntent menuItemPendingIntentTwo =
   createPendingEmailIntent();
intentBuilder.addMenuItem(menuItemEmailTitle,
   menuItemPendingIntentTwo);
```

Configuring custom enter and exit animations

Nothing is complete without a few animations to tag along. This is no different, as we have two transitions to make: one for the custom tab to enter and another for its exit; we have the option to set a specific animation for each start and exit animation:

```
//Setting start and exit animation for the custom tab.
intentBuilder.setStartAnimations(this,R.anim.slide_in_right,
   R.anim.slide_out_left);
intentBuilder.setExitAnimations(this,
   android.R.anim.slide_in_left, android.R.anim.slide_out_right);
```

Chrome warm-up

Normally, after we finish setting up the intent with the intent builder, we should call `CustomTabsIntent.launchUrl (Activity context, Uri url)`, which is a nonstatic method that will trigger a new custom tab activity to load the URL and show it in the custom tab. This can take up quite some time and impact the impression of smoothness the app provides.

We all know that users demand a near-instantaneous experience, so Chrome has a service that we can connect to and ask it to warm up the browser and its native components. Calling this will ask Chrome to perform the following:

- The DNS preresolution of the URL's main domain
- The DNS preresolution of the most likely subresources
- Preconnection to the destination, including HTTPS/TLS negotiation

The process to warm up Chrome is as follows:

1. Connect to the service.
2. Attach a navigation callback to get notified upon finishing the page load.

3. On the service, call `warmup` to start Chrome behind the scenes.

4. Create `newSession`; this session is used for all requests to the API.

5. Tell Chrome which pages the user is likely to load with `mayLaunchUrl`.

6. Launch the intent with the session ID generated in step 4.

Connecting to the Chrome service

Connecting to the Chrome service involves dealing with **Android Interface Definition Language (AIDL)**.

If you don't know about AIDL, read:

`http://developer.android.com/guide/components/aidl.html`

The interface is created with AIDL, and it automatically creates a proxy service class for us:

```
CustomTabsClient.bindCustomTabsService()
```

So, we check for the Chrome package name; in our sample project, we have a special method to check whether Chrome is present in all variations. After we set the package, we bind to the service and get a `CustomTabsClient` object that we can use until we're disconnected from the service:

```
pkgName - This is one of several options checking to see if we have a
version of Chrome installed can be one of the following
static final String STABLE_PACKAGE = "com.android.chrome";
static final String BETA_PACKAGE = "com.chrome.beta";
static final String DEV_PACKAGE = "com.chrome.dev";
static final String LOCAL_PACKAGE = "com.google.android.apps.chrome";

private CustomTabsClient mClient;

// Binds to the service.
CustomTabsClient.bindCustomTabsService(myContext, pkgName, new
  CustomTabsServiceConnection() {
  @Override
  public void onCustomTabsServiceConnected(ComponentName name,
    CustomTabsClient client) {
    // CustomTabsClient should now be valid to use
    mClient = client;
  }

  @Override
  public void onServiceDisconnected(ComponentName name) {
```

```
    // CustomTabsClient is no longer valid which also
      invalidates sessions.
      mClient = null;
    }
});
```

After we bind to the service, we can call the proper methods we need.

Warming up the browser process

The method for this is as follows:

```
boolean CustomTabsClient.warmup(long flags)

//With our valid client earlier we call the warmup method.
mClient.warmup(0);
```

 Flags are currently not being used, so we pass 0 for now.

The warm-up procedure loads native libraries and the browser process required to support custom tab browsing later on. This is asynchronous, and the return value indicates whether the request has been accepted or not. It returns true to indicate success.

Creating a new tab session

The method for this is as follows:

```
boolean CustomTabsClient.newSession(ICustomTabsCallback callback)
```

The new tab session is used as the grouping object tying the mayLaunchUrl call, the VIEW intent that we build, and the tab generated altogether. We can get a callback associated with the created session that would be passed for any consecutive mayLaunchUrl calls. This method returns CustomTabsSession when a session is created successfully; otherwise, it returns Null.

Setting the prefetching URL

The method for this is as follows:

```
boolean CustomTabsSession.mayLaunchUrl (Uri url, Bundle extras,
  List<Bundle> otherLikelyBundles)
```

This method will notify the browser that a navigation to this URL will happen soon. Make sure to `warmup()` prior to calling this method – this is a must. The most likely URL has to be specified first, and we can send an optional list of other likely URLs (`otherLikelyBundles`). The list have to be sorted in a descending order and the optional list may be ignored. A new call to this method will lower the priority of previous calls and can result in URLs not being prefetched. Boolean values inform us whether the operation has been completed successfully.

Custom tabs connection callback

The method for this is as follows:

```
void CustomTabsCallback.onNavigationEvent (int navigationEvent,
  Bundle extras)
```

We have a callback triggered upon each navigation event in the custom tab. The `int` `navigationEvent` element is one of the six that defines the state the page is in. Refer to the following code for more information:

```
//Sent when the tab has started loading a page.
public static final int NAVIGATION_STARTED = 1;
//Sent when the tab has finished loading a page.
public static final int NAVIGATION_FINISHED = 2;
//Sent when the tab couldn't finish loading due to a failure.
public static final int NAVIGATION_FAILED = 3;
//Sent when loading was aborted by a user action.
public static final int NAVIGATION_ABORTED = 4;
//Sent when the tab becomes visible.
public static final int TAB_SHOWN = 5;
//Sent when the tab becomes hidden.
public static final int TAB_HIDDEN = 6;
private static class NavigationCallback extends CustomTabsCallback {
  @Override
  public void onNavigationEvent(int navigationEvent, Bundle
    extras) {
    Log.i(TAG, "onNavigationEvent: Code = " + navigationEvent);
  }
}
```

Summary

We learned about a newly added feature, Chrome custom tabs, which allows us to embed web content into our application and modify the UI. Chrome custom tabs allow us to provide a fuller, faster in-app web experience for our users. We use the Chrome engine under the hood, which allows faster loading than regular WebViews or loading the entire Chrome (or another browser) application.

We saw that we can preload pages in the background, making it appear as if our data is blazing fast. We can customize the look and feel of our Chrome tab so that it matches our app. Among the changes we saw were the toolbar color, transition animations, and even the addition of custom actions to the toolbar.

Custom tabs also benefit from Chrome features such as saved passwords, autofill, tap to search, and sync; these are all available within a custom tab. For developers, integration is quite easy and requires only a few extra lines of code in the basic level. The support library helps with more complex integration, if required.

This is a Chrome feature, which means you get it on any Android device where the latest versions of Chrome are installed. Remember that the Chrome custom tab support library changes with new features and fixes, which is the same as other support libraries, so please update your version and make sure that you use the latest API to avoid any issues.

In our next chapter, we will take a deep breath and look at some of the new authentication/security features Android Marshmallow has to offer.

8
Authentication

Android Marshmallow has introduced a newly integrated API to better support user authentication and user verification. We can now use the new `Fingerprint` API for devices with a fingerprint scanner in order to authenticate the user. We can also set a specific time for user lock screen verification to be considered valid in the app login. In this chapter, we will try and go over these additions and explain how to use them:

- The `Fingerprint` authentication API
- Credentials' Grace Period
- `Cleartext` network traffic

The Fingerprint authentication API

Android Marshmallow now allows us, the developers, to authenticate users with their fingerprint scans when using such authentication scanners on supported devices.

The `Fingerprint` API was added to Android Marshmallow via a whole new package:

`android.hardware.fingerprint`

The package contains four classes:

- `FingerprintManager`
- `FingerprintManager.AuthenticationCallback`
- `FingerprintManager.AuthenticationResult`
- `FingerprintManager.CryptoObject`

Each class has a specific role in our fingerprint authentication process.

How do we use fingerprint authentication?

The preceding four classes of the `android.hardware.fingerprint` package can be explained in the following manner:

- `FingerprintManager`: Manage access to fingerprint hardware

- `FingerprintManager.AuthenticationCallback`: Callback used in the auth process

- `FingerprintManager.AuthenticationResult`: Result container for `auth` process

- `FingerprintManager.CryptoObject`: Specific `Crypto` object to use with `FingerprintManager`

Say, we want to authenticate users via their fingerprints. A device with a fingerprint sensor must be in use; otherwise, we can't use this API. We need to get an instance of `FingerprintManager`, and then we call the `authenticate()` method. We must implement a specific user interface for the fingerprint authentication flow, and the standard Android fingerprint icon (`c_fp_40px.png`) is included in the source. We need to add the appropriate permission to our app's manifest:

```
<uses-permission android:name="android.permission.USE_FINGERPRINT" />
```

Right now, we don't have a device with a fingerprint sensor, so we will need to test our code from an emulator. (Nexus 5X and Nexus 6P are still with limited supply)

Setting up for testing

Android SDK Tools Revision 24.3 (at least) must be installed. Now, we navigate to
Settings | Security | Fingerprint and add one fingerprint.

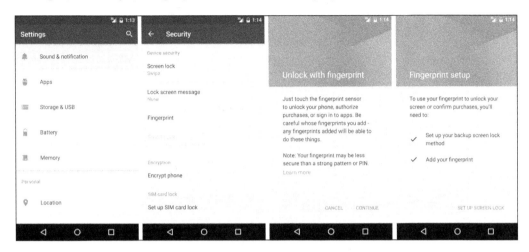

Follow the instructions manually; we are asked to select the PIN and leading us to
find the following screenshot:

Finally, we must use a special `adb` command, tricking the sensor into capturing
a mock fingerprint:

```
adb -e emu finger touch <finger_id>
```

The resultant screen should look like the following screenshot:

We used `finger_id =1` for a single finger. The same command also emulates fingerprint touch events on the lock screen or in our app.

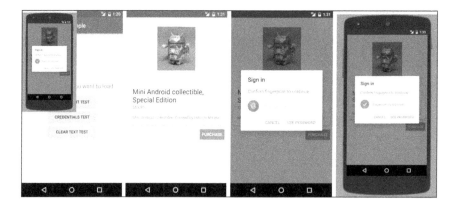

If you need help to set up an emulator, read:

`https://developer.android.com/tools/devices/index.html`

Now, we can launch our application and see that we can use the fingerprint as our authentication method when the user purchases an item.

Credentials' Grace Period

Ever got the itch when you wanted to use an app after device unlock only to find that you need to log in again or enter the app password again? Well, now we can query the device and check whether it was unlocked recently and how recent was it. This will give our users a chance to avoid all the fuss that comes with using our app. Note that this must be used in conjunction with a public or secret key implementation for user authentication. If you want to read more about the **Android Keystore System**, head to https://developer.android.com/training/articles/keystore.html.

We use `KeyguardManager` and check whether our lock screen is secured via the `isKeyguardSecure()` method. Once we know that it's secured, we can try and use the feature; otherwise, it'd imply that the user didn't set a secure lock screen and this feature is a *no-op*.

We generate a symmetric key with `KeyGenerator` in Android KeyStore, which can only be used after the user has authenticated with device credentials within the last x seconds. Setting this value (x) is done via the `setUserAuthenticationValidity DurationSeconds()` method, when we set up `KeyGenerator` or `KeyPairGenerator`.

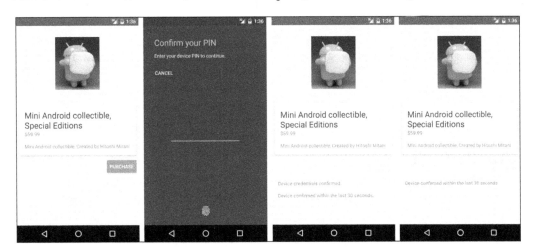

You can check out the sample code for more information. The activity is called `CredGraceActivity`.

 Try and display the reauthentication dialog as less as possible. When using a cryptographic object, you should try and verify its expiry, and only if it passes, use `createConfirmDeviceCredentialIntent()` to reauthenticate the user.

Cleartext network traffic

Android Marshmallow also added a new flag to the manifest. This flag indicates whether the application is using a `cleartext` network traffic such as HTTP. The flag is `android:usesCleartextTraffic`, and the default value is `true`. Setting this to `false` means that some system API components—such as HTTP and FTP stacks, `DownloadManager` and `MediaPlayer`—will refuse to issue HTTP traffic and will only allow HTTPS. It would be a good practice to build a third-party library that honor this setting as well. Why is this good? Well, `cleartext` traffic lacks confidentiality, authenticity, and protections against tampering, and data can be tempered without it being detected. This is a major risk for applications, and we can now use it to try and enforce a stronger and more secure data transport to/from our applications.

We need to remember that this flag is honored on the basis of the best effort, and it's not possible to prevent all `cleartext` traffic from Android applications given that they have permissions to use the `Socket` API, for instance, where the `Socket` API cannot determine `cleartext` usage. We can check out this flag by reading it from either `ApplicationInfo.flags` or `NetworkSecurityPolicy.isCleartextTrafficPermitted()`.

 WebView does not honor this flag, which means that it will load HTTP even if the flag is `false`.

So, what do we do with the cleartext network traffic flag?

During app development, we can use `StrictMode` and identify any `cleartext` traffic from our app using `StrictMode.VmPolicy.Builder.detectCleartextNetwork()`.

The downside of `usesCleartextTraffic` is that it causes app crashes or process termination when it's not using **SSL** (short for **Secure Socket Layer**). This is great in theory but not in production, where your SSL certificate, for some reason, has issues and you reroute the traffic to HTTP. So, pay extra attention to where HTTPS is used in your app and where it's okay to use HTTP.

Luckily, we have `StrictMode`, which now has a way to warn you if your application executes any unencrypted network operations via a `detectCleartextNetwork()` method on `StrictMode.VmPolicy.Builder`. In our sample project, we have a `ClearTextNetworkUsageActivity` activity; when running the `TestStrictHttp` `productFlavor` variant, you will see this in `LogCat`.

Summary

Android Marshmallow gave us a new API to authenticate users with the `Fingerprint` API. We can use the sensor, authenticate the user even within our application, and save it for later use if we want to save the need for user login using the Credentials' Grace Period capabilities Android Marshmallow introduced.

We also covered a way to make our application more secure using HTTPS only, and the `StrictMode` policy, enforced with the help of the `usesCleartextTraffic` flag, which allows us to make sure that all the nodes we connect to the outer world and examine the need for are a secure connection or not.

I would like to thank you for reading.

I would like to thank the Android team. This product has changed my life.

The Android ecosystem has great developers contributing by publishing libraries, writing blog posts and answering support questions; I'm proud to be part of it.

Looking forward for future editions.

Index

A

Android ART 53
Android Backup Service 30
Android Debug Bridge (adb) 26
Android for Work
  about 69
  URL 69
android.hardware.fingerprint package
  classes 89
  FingerprintManager 90
  FingerprintManager.Authentication
      Callback 90
  FingerprintManager.Authentication
      Result 90
  FingerprintManager.CryptoObject 90
Android Intent system
  about 21, 22
  app link settings and management 24
  app link verification, triggering 23, 24
  website association, creating 23
Android Interface Definition
      Language (AIDL)
  URL 85
Android Keystore changes 52
Android Keystore System
  URL 93
Android Marshmallow
  about 1, 61
  audio features 61
  behavioral changes 70
  camera features 65
  video features 64
Android Marshmallow changes
  apps, silently installing 73
  apps, silently uninstalling 73

  automatic system updates 73
  behavioral changes 70
  data usage statistics, querying 74
  improved certificate access, granting 73
  runtime permissions, managing 74
  single-use device improvements 72
  third-party certificate installation 74
  VPN, accessing 75
  VPN, configuring 75
  work profile status 75
Android Marshmallow permissions
  about 9
  app signature permissions granted 9
  best practices 13, 14
  permissions, declaring 9
  INTERNET permission 9
  managing 15, 16
  permission groups 9, 10
  permissions granted by users at runtime 9
  permissions, revoking 10
  PROTECTION_NORMAL permissions 9
  runtime permissions 10
Android permissions
  about 1-3
  permission group definitions 3, 4
  permissions, implied by feature
      requirements 4, 5
  viewing, for each app 5-8
Android SDK Tools Revision 24.3 90
Android Support Library 51
Android system flags permissions 17
Apache HTTP client removal 50
APK validation 54
app links
  Digital Asset Links API 25
  domains, listing 25

intent, testing 25
manifest, checking 25
policies, checking with adb 26, 27
testing 25

apps
installing, silently 73
uninstalling, silently 73

App Standby mode
about 43
apps, testing with 44
device, in App Standby mode 44
excluded apps and settings 45-48
points and tips 48, 49

assistant 58

Assist API 58

audio features, Android Marshmallow
about 61
audio and input devices 63
changes, in AudioManager 63
digital audio capture and playback 63
information on audio devices 63
MidiManager 62
support, for MIDI protocol 61, 62

authenticate() method 90

automatic backup
about 30
BackupAgent 36
backup events 37
excluded data 36
subtopics 35

automatic system updates 73

B

backup configuration syntax
<exclude> tag 32
<include> tag 32
domain 32

backup configuration testing
about 33
backup logs, setting 33
backup phase, testing 33
restore phase, testing 34
troubleshooting 34

Backup Manager service 30

behavioral changes, Android Marshmallow
DevicePolicyManager changes 71
Wi-Fi configuration lock 70
Wi-Fi configuration options 70
work contacts, display option 70
Work Policy Controller, adding 71

Bluetooth API
about 59
stylus support 59

Bring Your Own Device (BYOD) 69

C

camera features
about 65
changes, in camera service 66, 67
flashlight API 65
reprocessing API 66

Chrome custom tab
about 77
browser process, warming up 86
Chrome service, connecting 85, 86
connection callback 87
custom action button 83
custom enter animations, configuring 84
custom exit animations, configuring 84
customization options 78
custom menu, configuring 83, 84
custom UI 80-82
implementation guide 80
new tab session, creating 86
prefetching URL, setting 86, 87
tab interaction 80, 82
using 78-80
warming up 84
WebView 77, 78

Chromium project 78

ChubbyTabb
URL 81

cleartext network traffic
about 94
downside 94
using 94

coding permissions
about 11
coding, for runtime permissions 12, 13
testing 11

Context methods 49
CredGraceActivity 93
customization options, Chrome
    custom tab 78
CustomTabsHelper
  URL 80
custom URI scheme 22

D

Dalvik 53
data backup configuration
  about 30
  backup configuration syntax 32
  data, including or excluding 31
  opting out, from app data backup 33
data usage statistics
  querying 74
DevicePolicyManager
  URL 72
DevicePolicyManager, changes
  createAndInitializeUser() 71
  createUser() 71
  EXTRA_PROVISIONING_ADMIN_EX-
    TRAS_BUNDLE, using 72
  EXTRA_PROVISIONING_DEVICE_AD-
    MIN_PACKAGE_CHECKSUM 71
  EXTRA_PROVISIONING_DEVICE_AD-
    MIN_SIGNATURE_CHECKSUM 72
  EXTRA_PROVISIONING_RESET_PRO-
    TECTION_PARAMETERS 72
  keyguard restrictions, setting 71
  RESULT_CANCELED 72
  setCameraDisabled() 71
  setKeyguardDisabledFeatures() 71
  set of settings, disabling 72
  setScreenCaptureDisabled() method 71
  WIFI_DEVICE_OWNER_CONFIGS_LOCK-
    DOWN setting, enabling 72
Digital Asset Links API 25
digital audio capture and playback 63
Digital Signature Algorithm (DSA) 52
Direct Share
  about 54-56
  best practices 57

Doze mode
  about 40
  apps, testing with 41-43
  device, in dozing state 41

E

emulator
  setting up, URL 92

F

Fingerprint authentication API
  about 89
  adding 89
  setting up, for testing 90-92
  using 90

G

Google Cloud Messaging (GCM) 30
Google I/O 2015 58
Google Play for Work 73
Grace Period, Credentials
  using 93
Group ID 1

H

hardware identifier 53

I

improved Bluetooth low
    energy scanning 60
improved certificate access
  granting 73
intent filters 22
isKeyguardSecure() method 93

J

JSON file 23

K

kiosks 73

L

launch handler 22
Linux user ID 1

M

MidiManager 62
MIDI protocol 61

N

notifications feature 51

P

pending intent
    URL 82
permission failure 2
power-saving modes
    about 40
    App Standby mode 43
    Doze mode 40

R

removable storage adoption 49, 50
reprocessing API
    about 66
    android.media.ImageReader 66
    android.media.ImageWriter 66
runtime permissions
    managing 74

S

Secure Socket Layer (SSL) 94
sendBroadcast(Intent) method 2
setUserAuthenticationValidityDuration
    Seconds() method 93
SharedPreferences 37
single-use device improvements
    setKeyguardDisabled(), using 72
    setStatusBarDisabled(), using 72
    STAY_ON_WHILE_PLUGGED_IN,
        using 72
    UserManager.DISALLOW_SAFE_BOOT,
        using 72

stylus 59
SystemUpdatePolicy
    TYPE_INSTALL_AUTOMATIC option 73
    TYPE_INSTALL_WINDOWED option 73
    TYPE_POSTPONE option 73

T

text selection
    about 51
    support library notice 52
third-party certificate installation
    DevicePolicyManager APIs, calling 74
Torchi 65

U

URI (Uniform Resource Identifier) 22
USB connection 54

V

video features
    about 64
    android.media.MediaSync 64
    MediaCodecInfo.CodecCapabilities.get-
        MaxSupportedInstances 64
    MediaPlayer.setPlaybackParams 65
voice actions 58
voice interactions 58
VPN
    device owner 75
    profile owner 75

W

WebView 77, 78
Wi-Fi and networking APIs 52
Work Policy Controller
    about 71
    URL 71
work profile status
    checking 75

Y

YUV 66

Thank you for buying
Android 6 Essentials

About Packt Publishing

Packt, pronounced 'packed', published its first book, *Mastering phpMyAdmin for Effective MySQL Management*, in April 2004, and subsequently continued to specialize in publishing highly focused books on specific technologies and solutions.

Our books and publications share the experiences of your fellow IT professionals in adapting and customizing today's systems, applications, and frameworks. Our solution-based books give you the knowledge and power to customize the software and technologies you're using to get the job done. Packt books are more specific and less general than the IT books you have seen in the past. Our unique business model allows us to bring you more focused information, giving you more of what you need to know, and less of what you don't.

Packt is a modern yet unique publishing company that focuses on producing quality, cutting-edge books for communities of developers, administrators, and newbies alike. For more information, please visit our website at www.packtpub.com.

About Packt Open Source

In 2010, Packt launched two new brands, Packt Open Source and Packt Enterprise, in order to continue its focus on specialization. This book is part of the Packt Open Source brand, home to books published on software built around open source licenses, and offering information to anybody from advanced developers to budding web designers. The Open Source brand also runs Packt's Open Source Royalty Scheme, by which Packt gives a royalty to each open source project about whose software a book is sold.

Writing for Packt

We welcome all inquiries from people who are interested in authoring. Book proposals should be sent to author@packtpub.com. If your book idea is still at an early stage and you would like to discuss it first before writing a formal book proposal, then please contact us; one of our commissioning editors will get in touch with you.

We're not just looking for published authors; if you have strong technical skills but no writing experience, our experienced editors can help you develop a writing career, or simply get some additional reward for your expertise.

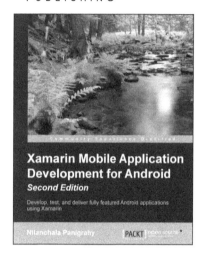

www.ingramcontent.com/pod-product-compliance
Lightning Source LLC
LaVergne TN
LVHW081346050326
832903LV00024B/1348